Nuclear Astrophysics A Course of Lectures

Nuclear Astrophysics- A Course of Lectures

Dr. Md A Khan

Department of Physics,
Aliah University,
New Town, Kolkata

CRC Press
Taylor & Francis Group
Boca Raton London New York

CRC Press is an imprint of the
Taylor & Francis Group, an **informa** business

CRC Press
Taylor & Francis Group
6000 Broken Sound Parkway NW, Suite 300
Boca Raton, FL 33487-2742

First issued in paperback 2023

ISBN-13: 978-1-138-58816-5 (hbk)
ISBN-13: 978-1-03-265336-5 (pbk)
ISBN-13: 978-0-429-49245-7 (ebk)

DOI: 10.1201/9780429492457

Print edition not for sale in South Asia (India, Sri Lanka, Nepal, Bangladesh, Pakistan or Bhutan)

Publisher's Note
The publisher has gone to great lengths to ensure the quality of this reprint but points out that some imperfections in the original copies may be apparent.

Library of Congress Cataloging in Publication Data
A catalog record has been requested

Visit the Taylor & Francis Web site at
http://www.taylorandfrancis.com

and the CRC Press Web site at
http://www.crcpress.com

LEVANT

Preface

Nuclear Astrophysics is being taught as a part of the Post Graduate Curriculum in Physics in majority of Indian Universities and Institutes. This book is largely based on the lectures delivered to postgraduate students in Physics of the Aliah University for the past few years. The materials are presented in some thirty odd lectures and I believe there is much in them for a good one semester course. However this work is intended for both the undergraduate and post graduate students interested in Nuclear Astrophysics, and also for a more general audience since no specific expertise in astronomy, physics and mathematics is required to fruitfully read majority of the text presented in this book.

This book contains 11 chapters. The opening Chapter 1 started with a brief introduction to nuclear astrophysics followed by discussion on elementary concepts of nuclear, particle and astrophysics. In chapter 2 nucleosynthesis and its different stages are discussed. Chapter 3 devoted in the description of the reaction cycles starting from hydrogen burning stages to s- and r-neutron processes. In chapter 4 an attempt has been made to explore the opportunities in nuclear astrophysics from both theoretical and experimental points of views. In chapter 5 explosive burning processes responsible for the nucleosynthesis of chemical elements are discussed. In chapter 6 the core collapse supernovae along with some relevant phenomenon have in explained. In chapter 7 the basic components of r-processes such as - the role of nuclear physics in the r-process, required conditions for r-process, r-process sites, and neutron star mergers are discussed. Phenomenon like nova explosions, X-ray bursts including ideas of black hole and neutron stars are discussed in chapter 8 in the context of nuclear processes in explosive binary star systems. In chapter 9 elementary ideas of the formation of stars and galaxies are introduced. In chapter 10 computational aspects of nuclear astrophysics have been discussed in context of stellar structure and evolution, radiation transfer and some other relevant phenomenon's. In chapter 11 an account of the major challenges and future prospects of nuclear astrophysics have been discussed. Important bibliographic references for further studies have been added at the end of each chapter.

I am thankful to many of my students who have been attending my lectures and especially to Mahamadun Hassan who constantly insisted me to shape the material presented here. I also acknowledge my gratitude to Prof. J. Alam and Dr D N Basu of VECC Kolkata for extending their support by opening avenue for me to do research collaboration in Nuclear Astrophysics. Thanks are due to all my colleagues, friends and staff at the Aliah University for their support and encouragements.

Last but not least, I would like to thank my wife, daughters Isra, Sara and son Rayyan for their patience and devotion.

Finally I would like to thank the entire team and especially to Mr Millinda Dey of Sarat Book House, Kolkata for their effort to bring the materials in the present form.

All suggestions and criticism for the improvement of this edition will be cordially received.

May 9, 2016

New Town, Kolkata, India   Dr Md Abdul Khan

Contents

1 Basics of Nuclear Astrophysics   11
   1.1   Introduction 11
   1.2   Nuclear Physics 16
      1.2.1   Branching of Radioactivity 17
      1.2.2   Binding energy 18
      1.2.3   Nuclear Reactions 21
      1.2.4   Reaction Q-value 22
      1.2.5 α decay 23
      1.2.6 β-decay and electron capture 25
      1.2.7 γ-decay 27
      1.2.8   Nuclear fusion 28
      1.2.9   The Coulomb barrier 29
      1.2.10   Reaction rate in a medium 36
      1.2.11   Compound nuclear hypothesis 38
      1.2.12   Breit-Wigners one-level formula 41
      1.2.13   Cross-sections and their derivations 42
   1.3   Particle physics 47
   1.4   Astrophysics 50
   1.5   Scopes of nuclear astrophysics 51

2 Nucleosynthesis   61
   2.1   Introduction 61
   2.2   Primordial nucleosynthesis 67
   2.3   Stellar nucleosynthesis up to iron ($A \leq 60$) 73
   2.4   Stellar nucleosynthesis beyond iron(A>60) 81

3 Sources of Energy in Stars   93
   3.1   Introduction 93
   3.2   Hydrogen burning and p-p (proton-proton) chain 95

3.3   CNO Cycles   97

3.4   The Hot CNO Cycle and the Rapid Proton Process   101

3.5   The Early rp–Process and Bottleneck Reactions.   103

3.6   Helium (^{4}He) Burning   105

3.7   The Triple-α Process   105

3.8   Heavier Element α-Burning   107

    3.8.1   Neutron Production in (α, n) Reactions.   107

3.9   The s and r-Neutron Processes.   107

4   Opportunities in Nuclear Astrophysics   **113**

4.1   Introduction   113

4.2   Thermonuclear rates and reaction networks   114

    4.2.1   Thermonuclear Reaction Rates   114

    4.2.2   Nuclear Reaction Networks   117

    4.2.3   Burning Processes in Stellar Environments   119

4.3   Experimental Techniques in Nuclear Astrophysics   122

    4.3.1   Choice of Target   123

    4.3.2   Choice of detectors   127

    4.3.3   Recoil separators   128

    4.3.4   Ground State Properties   128

    4.3.5   Resonances Properties   131

    4.3.6   Transfer Reactions   132

4.4   Experimental Nuclear Astrophysics   134

    4.4.1   Energy ranges for measurement of cross-sections . . .   134

    4.4.2   Radioactive Beams   135

    4.4.3   In-Flight Separators.   138

4.5   Cross Section Predictions and Reaction Rates   140

    4.5.1   Thermonuclear Rates from Statistical Model Calcula-
        tions   142

    4.5.2   Astrophysical S-factors of radiative capture reactions .   152

    4.5.3   Sub-barrier fusion and selective resonant tunneling cross
        section   156

4.6   Weak-Interaction Rates   165

    4.6.1   Electron Capture and β-Decay   166

    4.6.2   Neutrino-Induced Reactions   169

5   Explosive Burning Processes   **177**

5.1   Introduction   177

5.2   Explosive H-Burning   179

5.3   Explosive He-Burning   183

5.4 Explosive C- and Ne-Burning 185
5.5 Explosive O-Burning 185
5.6 Explosive Si-Burning 186
5.7 The r-Process 189

6 Core Collapse Supernovae 193
6.1 Introduction 193
6.2 General Scenario 198
6.3 Role of weak-Interaction rates in presupernova evolution . . . 200
6.4 The Role of Electron Capture During Collapse 202
6.5 Neutrino-Induced Processes During a Supernova Collapse . . 208
6.6 Type II Supernovae Nucleosynthesis 209

7 Basic Components of Astrophysical r-Process 219
7.1 Role of Nuclear Physics in the r-process 220
7.2 Parameters controlling the r-process 223
7.3 r-Process Sites 224
  7.3.1 Type II Supernovae. 224
  7.3.2 Neutron Star Mergers 226
  7.3.3 r-Process Overview 226

8 Nuclear Processes in Explosive Binary Stars 231
8.1 Introduction 231
8.2 Nova Explosions 232
8.3 X-Ray Bursts 234
8.4 Thermonuclear Runaway. 236
8.5 The rp-Process. 239
8.6 X-Ray Pulsars 241
8.7 Accretion Processes on Neutron Stars and Black holes 241

9 Formation of stars and galaxies 243
9.1 Introduction 243
9.2 Life cycle of Stars 244

10 Role of Computation in Nuclear Astrophysics 255
10.1 Introduction 255
10.2 Computational versus Analytic Methods 259
10.3 Major Areas of Application of CNA 259
  10.3.1 Stellar structure and evolution 260
  10.3.2 Radiation transfer and stellar atmospheres 260
  10.3.3 Astrophysical fluid dynamics 260

10.3.4   Planetary, stellar and galactic-dynamics 260

10.3.5   Numerical Methods: 261

10.3.6   Stellar Structure Codes 261

10.3.7   Radiative Transfer Codes 261

10.3.8   N-body Codes 262

10.3.9   Codes for Astrophysical Fluid Dynamics 262

10.3.10 Relation to other fields: 262

10.3.11 Numerical Analysis 262

10.3.12 Computer Science 264

10.3.13 Relevant History 264

11 Key challenges and future scopes   267

11.1   Introduction 267

11.2   Key challenges 269

11.3   Demand for infrastructure expansion 270

11.4   Progress Scenario 272

11.4.1   Available Data Resources 275

11.4.2   Future Data Developments 276

11.5   Present and Future Projects & their Benefits 279

11.5.1   Astronomical observatories 279

11.5.2   Benefit to theory and computation 280

11.5.3   Benefit to the society 281

Index   283

Chapter 1

Basics of Nuclear Astrophysics

1.1 Introduction

Physics is a resourceful subject having many branches and with the advancement of time many new ideas, observations, theories, and experimental facts still adding newer branches to it. Nuclear astrophysics is the study of nuclear-level processes that occur naturally in space. Notably, this includes understanding the chain of fusion events, or nucleosynthesis , that occurs in stars, and how this can be detected from a distance by measuring the radiation these processes produce. Thus the hybridization of the ideas of astrophysics with those of nuclear physics led to the development of this new branch of physics which we call as nuclear astrophysics. It is an interdisciplinary branch of physics involving close collaboration among researchers in various subfields of nuclear physics and astrophysics, with significant emphasis in areas such as stellar modeling, measurement and theoretical estimation of nuclear reaction rates in the stellar environment, cosmology, cosmochemistry, gamma-ray, optical and X-ray astronomy, advancing our knowledge towards nuclear lifetimes and masses. Generally, nuclear astrophysics aims i) to understand the origin of the chemical elements and production of energy in stars by focusing on the explosive stellar events (known as supernova) and the associated nuclear phenomena, ii) to construct theoretical models for supernovae, nova, x-ray, and gamma-ray bursts, iii) to understand the creation of new atomic nuclei and their contribution in the formation of galaxies, new stars and planets, and iv) to explore the possible way to explain how the study of the stellar events and their nuclear products, that

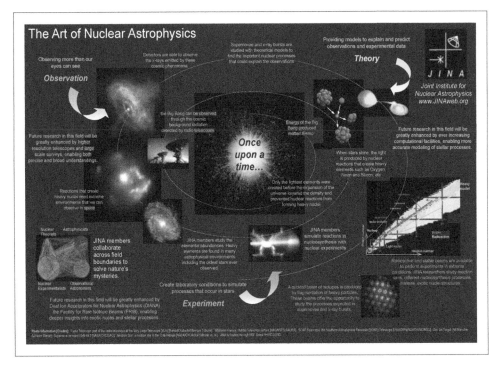

Figure 1.1: The Art of Nuclear Astrophysics [Source: www.JINAweb.org].

form much of ourselves and our world, could connect us to the beginning of the Universe. Figure 1.1 tells about the major subject areas of nuclear astrophysics. The environment where we live in always triggers us in many ways so that we may understand it. For instance, when we are exposed to sun rays in hot summer days we are being invited by the nature to learn about the source of the radiant energy of the sun, which is far away from our earth. In this way, nature itself has been insisting us to search for the answer of many fundamental questions: How the present Universe came into existence? What existed at the beginning of space and time? How elements are synthesized? How stars are formed? How galaxies are formed? Why stars shine? How stars die? How supernova works? How neutron stars are produced? How black holes or indexwarm holewarm holes are formed? How old is the universe? And the list continues. Since the inception of modern science, brilliant mathematicians and scientists have sought the answers to these fundamental questions. Among those Copernicus, Galileo, Newton, Einstein, Hubble, and others used direct observation, reasoning, used mathematics, and new technologies to overturn ideas about cosmology that were believed to be fundamental truths. Their breakthroughs reshaped the scientific understanding of the nature and structure of the universe. Their work, together with works of other well-known cosmologists, not only provided new explanations of the underlying facts about the universe, but also raised seemingly ironic questions. Did the enormous variety and mass of matter that make up the cosmos evolve from nothing but energy? If that is true, then where did the energy that produced all of the matter in the universe come from? A literature survey on the issue reveals that: the basic principles of explaining the origin of the elements and the energy generation in stars were laid down in the theory of nucleosynthesis which came together in the late 1950s from the seminal works of Burbidge, Burbidge, Fowler and Hoyle and independently by Cameron. Among them, Fowler is largely credited for initiating the collaboration among astronomers, astrophysicists, and experimental nuclear physicists to synthesize the subject which is what we now know as nuclear astrophysics. He won the Nobel Prize in 1983 for his remarkable contribution in nuclear astrophysics. The basic requirement of experimental nuclear and particle physicists is the projectile beam of appropriate energy and intensity together with appropriate well maintained reaction/event friendly environment. To some extent the required beam can be made available by artificially designed particle accelerators and reactors. However, to nuclear and particle physicists, the early universe represents the ultimate particle accelerator as well as nuclear reactor in which energy density and particle density are both beyond the values expected to achieve in artificially designed accelerators and reactors on the planet we are residing. Although in the early universe reactions occurred with incomprehensible rates and varieties, still there is hope, to successfully explain many aspects of those reactions including their underlying processes by careful study of their end products. Above all, the most challenging job one has to encounter in the study of nuclear astrophysics is the understanding of the mechanism of formation of heavy elements in the heart of the stars by fusion and neutron capture reactions. This is because- the nuclear physics, the mechanics and thermodynamics involved in such environment is very complicated to be understood. The principal source of stellar informatory evidences are the data accumulated by astronomical observations with the latest equipments like γ-ray spectrometers carried in orbiting space vehicles in addition to the conventional optical telescopes. The golden age of nuclear physics started accidentally in 1896 with the discovery of natural radioactivity by Henri Becquerel by his famous observation of the event of blackening of the photographic plates kept in the vicinity of uranium-sulfide crystals which brought to him the Nobel Prize for Physics in 1903. Further advancement of the understanding of many characteristics of radio-active substances was put forward by Pierre and Marie Curie by chemical isolation of different radioactive elements produced in the decay processes of uranium. In 1899 Ernest Rutherford performed a number of brilliant experiments leading to his discovery of atomic nucleus in 1911. In this chain of knowledge some points should also been reserved for Tycho Brahe and his fellow Johannes Kepler who two observed and studied radioactive phenomena in bright stellae novae, i.e. new stars in 1572 (Brahe) and in 1603 (Kepler). Such supernovae are now believed to be the explosions of aged stars at the termination of their normal lives. The post-explosion energy source of supernovae is the decay of radioactive nickel (^{56}Ni, half-life 6.077 days) and then cobalt (^{56}Co, half-life 77.27 days). Brahe and Kepler observed that the luminosity of their supernovae decreased with time at rate that could have been determined by the nuclear lifetimes. Like Becquerel, Brahe and Kepler did not realize the importance of what they had seen. In fact, the importance of supernovae dwarfs and that of radioactivity are much precious because they are the culminating events of the process of nucleosynthesis. This process started in the cosmological "big bang" where protons and neutrons present in the primordial soup condense to form hydrogen and helium. Later, when stars are formed, the hydrogen and helium are processed through nuclear reactions into heavier elements. These elements are then ejected into the interstellar medium by supernovae. Later, some of these matters condense to form new stellar systems, sometimes containing habitable planets made of the products of stellar nucleosynthesis. Nuclear physics has allowed us to understand the mechanisms of the processes by which elements are formed and also to determine their relative abundances. The distribution of nuclear abundances in the Solar System reveals the fact that most of the ordinary matter is in the form of hydrogen ($\sim 3/4$ by mass) and helium ($\sim 1/4$). Very small fraction ($\sim 1/50$) of the solar system material is in the form of heavy elements, especially carbon, oxygen and iron. The unknown cosmological "dark matter" lives near a hydrogen burning star and are made primarily of elements like hydrogen, carbon and oxygen. The existing theory of nucleosynthesis yields some particularly mesmerizing results that the observed mix of elements is due to a number of slight inequalities of nuclear and particle physics. These inequalities include the fact that i) the neutron is slightly heavier than the proton; ii) the neutron–proton system has only one bound state while the neutron– neutron and proton–proton systems are unbound; iii) the ^{8}Be nucleus is slightly heavier than gross of two ^{4}He nuclei and the second excited state of ^{12}C is slightly heavier than sum of three ^{4}He nuclei. Modification of any of these conditions would result in a radically

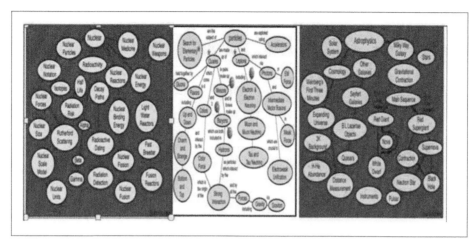

Figure 1.2: The chain connections among the basic elements of nuclear, particle and astrophysics. [Image source: www.google.com]

different distribution of elements. For instance, making the proton heavier than the neutron would make ordinary hydrogen unstable and none would have survived in the primitive age of the Universe. The extreme sensitivity of nucleosynthesis to nuclear masses has generated a considerable amount of debate about its interpretation. It hinges upon whether nuclear masses are fixed by the fundamental laws of physics or are accidental, perhaps taking on different values in inaccessible regions of the Universe. Nuclear mass depends on the strengths of the forces between neutrons and protons, and yet we do not know whether the strengths are uniquely determined by fundamental physics. If they are not, we must consider the possibility that the masses in "our part of the Universe" are as observed because other masses give mixes of elements that are less likely to provide environments leading to intelligent observers. Whether or not such "weak-anthropic selection" had a role in determining the observed laws of nuclear and particle physics has a question that is appealing to some and irritating to others. Resolving the question will require better understanding of the origin of observed physical laws. The chain connections among the basic elements of nuclear, particle and astrophysics are summarized in Figure 1.2. An illustration of different scales for measurements of the dimensions of natural objects are illustrated in Figure 1.3. In the following sections brief introduction of some of the basic features of nuclear, particle and astrophysics are included for better understanding of the subject.

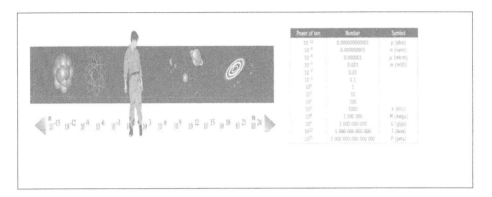

Figure 1.3: Scales and numbers relevant to nuclear astrophysics[Image source: CERN-Brochure-2008-001-Eng].

1.2   Nuclear Physics

Atomic nuclei are formed of two building blocks: neutrons and protons collectively referred as nucleons. Protons carry a single unit of positive charge ($e = 1.60 \times 10^{-19}C$), whilst neutrons are electrically neutral having zero charge. The masses of the two types of particle are similar, with the neutron being slightly more massive than the proton: $m_p = 1.6726 \times 10^{-27}kg$, $m_n = 1.6749 \times 10^{-27}kg$. An atomic nucleus is composed of Z protons and N neutrons held together by the strong nuclear force in a region of very small radius $r \simeq (1.21fm)A^{1/3}$ where A (= Z + N) is the mass number of the concerned nucleus. Each value of the atomic number Z bears the identity of a particular chemical element and also indicates the number of electrons in a neutral atom of that element. A particular nuclear species is specified by a symbol $\binom{A}{Z}Y_N$, or in the non-redundant form ^AY, where Y denotes the chemical symbol of the element. Atoms can transform from one type to another through processes like radioactive decays or nuclear fusion processes obeying certain conservation principles such that Electric charge is always conserved: the net charge of the products of a nuclear decay is the same as the net charge of the original nucleus. The mass number is conserved: the total number of nucleons in the products is the same as that in the original nucleus. The linear momentum is conserved: the total linear momentum of the products is the same as that of the initial particle(s). The angular momentum is conserved: the total angular momentum of the products is the same as that of the initial particle(s). The total energy including the rest-mass energy is conserved: in all nuclear processes the total energy, including the rest-mass energies of all the particles taking part in the nuclear processes must be equal to the sum of the kinetic energies and the rest-mass energies of the products. Since the energies involved in nuclear decays are large enough, we need to take account of the relationship between energy and mass, $E = mc^2$. As nuclear energies are generally measured in unit of MeV (mega electron volt) or GeV (giga electron volt), convenient units in which nuclear masses are to be measured are MeV/c^2 or GeV/c^2 or even larger TeV/c^2. In these units, the mass of a proton is $938.3MeV/c^2$ and that of a neutron is $939.6MeV/c^2$, or around $1GeV/c^2$ in each case. The phenomena of radioactive decay is governed by the exponential decay law

$$N(t) = N_0 e^{-\lambda t} \tag{1.1}$$

where $\lambda = \log_e(2)/T_{1/2} = 0.693/T_{1/2}$, N_0 is the initial number of radioactive nuclei and N is the average number left at time t. The half-life, $T_{1/2}$, characterizes the nucleus and its decay mode. The number (or mass) of active nuclei is halved during a period equal to one half-life. Again since radioactivity (A) is a measure of the intensity of radiation emitted by a radioactive material, which depends on the rate of transformation of the radioactive atoms, we may write

$$
\begin{aligned}
A &= |dN/dt| \\
&= \lambda N \\
&= \lambda N_0 e^{-\lambda t} \\
&= A_0 e^{-\lambda t} \tag{1.2}
\end{aligned}
$$

1.2.1 Branching of Radioactivity

In general radioactive elements undergo transformation either by α decay or β decay. In few instances both α and β disintegrations are observed for the same isotope with definite branching ratios for the α and β branches. If a radioactive material exhibits both α and β activities, then there are definite probabilities of decay in the two branches. Let λ_α and λ_β be the decay constants for the two branches. Then the probability of decay of the atom by both α and β emissions in time dt is $(\lambda_\alpha + \lambda_\beta)dt$. For N atoms, total rate of disintegration will be

$$
\begin{aligned}
\frac{dN}{dt} &= -(\lambda_\alpha + \lambda_\beta)N \\
&= -\lambda N \tag{1.3}
\end{aligned}
$$

where $\lambda = \lambda_\alpha + \lambda_\beta$, is the total decay constant. If the partial half-lives of disintegration by α and β emissions are $T_{1/2}^\alpha$ and $T_{1/2}^\beta$ respectively, then

$$\lambda_\alpha = \frac{ln2}{T_{1/2}^\alpha} \; and \; \lambda_\beta = \frac{ln2}{T_{1/2}^\beta}$$

and for the mean half-life $T_{1/2}$ of the substance, we can write

$$\lambda = \frac{ln2}{T_{1/2}}$$

. Hence we get

$$\frac{ln2}{T_{1/2}} = \frac{ln2}{T_{1/2}^\alpha} + \frac{ln2}{T_{1/2}^\beta}$$

or

$$\frac{1}{T_{1/2}} = \frac{1}{T_{1/2}^\alpha} + \frac{1}{T_{1/2}^\beta} \tag{1.4}$$

the ratios $\frac{\lambda_\alpha}{\lambda}$ and $\frac{\lambda_\beta}{\lambda}$ give the branching ratios for the two branches.

1.2.2   Binding energy

The binding energy B of a nucleus is the minimum energy required for dissociating the nucleus into its constituent nucleons; it is also equal to the energy released when the nucleus is formed from its constituents. The mass of an atom is found to be less than the total mass of its constituent neutrons (s) and proton(s) by an amount equal to the mass equivalent of B which in symbols can be represented as

$$M(A, Z) = Zm_H + (A - Z)m_n - B/c^2 \tag{1.5}$$

where M(A, Z), m_H and mn respectively denote the atomic masses of- the nuclide of mass number A and atomic number Z, a hydrogen atom and a neutron and B is expressed in energy unit. Thus one may note that: Nuclei are made up of protons and neutrons, but the mass of a nucleus is always less than the sum of the individual masses of the protons and neutrons which constitute it. The difference is a measure of the nuclear binding energy which holds the nucleons together. This binding energy is actually released during nuclear fusion reaction. For ^{4}He:
mass of two protons = $2 \times 1.00728u$, mass of two neutrons = $2 \times 1.00728u$, mass of four nucleons = 4.03188 u, mass of alpha = 4.00153 u, $1u = 1.66054\times$

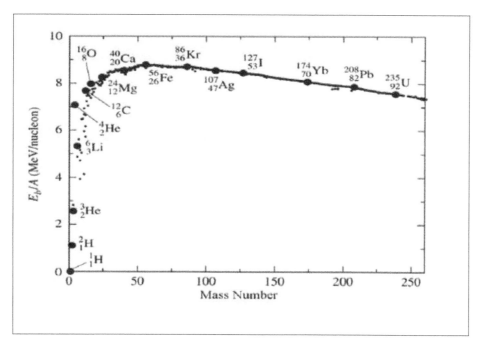

Figure 1.4: The binding energy per nucleon (B/A) as a function of mass number (A).

$10^{-27} kg = 931.494 MeV/c^2$

Binding energy (B) can be calculated using the relation, $B = \Delta m c^2$

For alpha (^{4}He):

$$\Delta m = 0.03035u, B = 0.03035 \times 931.494 MeV \simeq 28.3 MeV$$

It is important to note that mass of nuclei heavier than iron (Fe) is larger than the total mass of nuclei merged to form it. The binding energy per nucleon (B/A) gives a good guide (See Figure 1.4) to explain the nuclear stability or to know whether energy will be released in a nuclear fusion reaction or to explain why light elements undergo nuclear fusion reaction while the heavier undergoes decays or fissions. The most stable element ^{56}Fe has the highest B/A (=8.79 MeV/A). In the fusion of two nuclei having mass numbers A_1 and A_2, there will be a net gain in energy when $A_1 + A_2 < 56$ that will be released, but if $A_1 + A_2 > 56$, it costs energy to fuse the nuclei.

Table 1.1: Examples of binding energies for some nuclei.

Nucleus	Binding Energy, B (MeV)	Average binding energy per nucleon, B/A
^{2}H	2.22	1.11
^{4}He	28.30	7.07
^{12}C	92.16	7.68
^{40}Ca	342.05	8.55
^{56}Fe	492.26	8.79
^{238}U	1801.70	7.57

Seemi-empirical formulae

i) The Bethe-Weizsäcker semi-empirical formula for the binding energy of an isotope of ass number A, atomic number Z is written as

$$
\begin{aligned}
B(A, Z) &= E_v - E_s - E_c - E_a + E_p \delta(A, Z) \\
&= a_v A - a_s A^{2/3} - a_c Z^2/A^{1/3} \\
&\quad -a_{sym}(A - 2Z)^2/A + a_p A^{-3/4}\delta(A, Z) \qquad (1.6)
\end{aligned}
$$

ii) The Bethe-Weizsäcker semi-empirical formula for the atomic mass of an isotope of mass number A, atomic number Z is written as

$$
\begin{aligned}
M(A, Z) &= ZM_H + NM_n - B(A, Z) \\
&= ZM_H + (A - Z)M_n - a_v A + a_s A^{2/3} + a_c Z^2/A^{1/3} \\
&\quad +a_{sym}(A - 2Z)^2/A - a_p A^{-3/4}\delta(A, Z) \qquad (1.7)
\end{aligned}
$$

In the above equations the terms $E_v, E_s, E_c, E_{asy}, E_p$ indicate volume energy, surface energy, Coulomb energy, asymetry energy respectively.The coefficients $a_v, a_s, a_c, a_{asy}, a_p$ stand for the volume energy, surface energy, Coulomb energy, asymetry energy (arising due to isospin asymetry $T_z = |\frac{N-Z}{2}|$) and pairing energy terms respectively. For even A, even Z nuclei $\delta = +1$; for odd A, odd Z nuclei $\delta = -1$ and for odd A nuclei $\delta = 0$. The surface energy term dominates in the low mass region and looses its significance with increasing mass number A. The Coulomb effect dominates in the heavier nuclei having large number of protons region but it has less contribution in the low mass region. The asymetry energy term also has less significance in the low mass region where the difference N-Z (=A-2Z) is small but it becomes very significant in the higher mass region.

A set of presently accepted values of the coefficients in atomic mass unit are: a_v=0.096919u, a_s=0.019114u, a_c=0.0007626u, a_{asy}=0.02544u, a_p=0.036u, 1u=931.5MeV

1.2.3 Nuclear Reactions

When two nuclei collide there are 2 types of reactions: In the first type of reactions the colliding nuclei can coalesce to form highly excited Compound nucleus (CN) that lives for relatively long time, lifetime sufficient for excitation[1] energy to be shared by all nucleons. If sufficient energy is localized on one or more nucleons (usually neutrons) they can escape resulting in decays of CN. In this type of reactions the CN lives long enough and it loses its memory of how it was formed. So probability of various decay modes independent of entrance channel are prevalent in this type of reactions. In the second type of reaction in which the colliding nuclei make 'glancing' contact and separate immediately, are said to undergo Direct reactions (DI). In this type of reactions the projectile may lose some energy, or have one or more nucleons transferred to or from it. So, there are two types of reactions (1) direct reactions and (2) compound nuclear reactions.

Direct reactions

- Elastic scattering : A(a; a)A {zero Q-value and internal states unchanged}.

- Inelastic scattering: $A(a; a')A^*$ or $A(a; a^*)A^*$. Projectile a gives up some of its energy to excite target nucleus A. If the nucleus a is also a complex nucleus, it can also be excited. [If energy resolution in detection is not small enough to resolve g.s. of target from low-lying

---

[1]Excitation, in physics, the addition of a discrete amount of energy (called excitation energy) to a system- such as an atomic nucleus, an atom, or a molecule- that results in its alteration, ordinarily from the condition of lowest energy (ground state) to one of higher energy (excited state). In nuclear, atomic, and molecular systems, the excited states are not continuously distributed but have only certain discrete energy values. Thus, external energy (excitation energy) can be absorbed only in correspondingly discrete amounts. Thus, in a hydrogen atom (composed of an orbiting electron bound to a nucleus of one proton), an excitation energy of 10.2 electron volts is required to promote the electron from its ground state to the first excited state. A different excitation energy (12.1 electron volts) is needed to raise the electron from its ground state to the second excited state.

excited states then cross section will be sum of elastic and inelastic components. This is called quasi-elastic scattering].

- Breakup reactions: Usually referring to breakup of projectile a into two or more fragments. This may be elastic breakup or inelastic breakup depending on whether target remains in ground state.

- Transfer reactions: Stripping and pickup reactions.

- Charge exchange reactions: mass numbers remain the same and they can be elastic or inelastic.

Compound nuclear reactions

- Fusion: Nuclei stick together

- Fusion-evaporation: fusion followed by particle-evaporation and/or gamma emission

- Fusion-fission: fusion followed by fission

1.2.4   Reaction Q-value

The Q-value for reaction is given by the total mass energy before a reaction minus the total mass- energy after the reaction. Symbolically, a nuclear reaction may be represented as $A + B \rightarrow C + D$ or $B(A, C)D$, where A=projectile nucleus, B=target nucleus and C and D are the residual nuclei, A and B are described as the entrance channel and C and D are known as the exit channel. The Q-value for a reaction is given by

$$Q = (M_A + M_B - M_C - M_D)c^2 \qquad (1.8)$$

The atomic masses M_x are usually found in tabular form in terms of the mass excess,

$$\Delta m = (M - AM_u)c^2, \qquad (1.9)$$

where M=atomic mass, A=mass number, $M_u = 1$ atomic mass unit (AMU) = 931.5 MeV/c^2. For $Q > 0$ the reaction is said to be exothermic , for Q < 0 it is said to be endothermic. As described above the binding energy per nucleon, B/A, is a measure of the relative stability of nuclei. B/A for most nuclei is within 10% or so of 8 MeV per nucleon, with a maximum value of about 8.79 MeV per nucleon near A = 56 (iron and nickel), then falling off slowly for heavier nuclei and falling off quite rapidly for the light nuclei. This means that energy can be released when heavy nuclei undergo fission or when light nuclei undergo fusion. When the values of the total energy per nucleon (-B/A) for all known nuclei are plotted as points above the ZN plane they lie near a surface having the shape of a valley called the valley of stability , with the valley floor lying directly above the path of stability. Then the β^- decay of neutron-rich isobars and the β^+-decay (or electron capture) of proton-rich isobars can be thought of as streams running down the valley walls, while α-decay of heavy nuclei is like a stream running down the valley floor. The semi-empirical model of the nucleus incorporates the short-range strong nuclear force and the repulsive Coulomb force and is based on an analogy between a nucleus and a charged liquid drop. It identifies four contributions to B/A: a volume energy, a surface energy, a Coulomb energy and a symmetry energy. Using empirical parameters, the model fits the overall trend of measured B/A values quite well but there are discrepancies in the small-scale structure suggestive of shell effects. In the nuclear shell model each proton and neutron is represented by a wave function and occupies an energy level in a potential energy well produced by the other nucleons. The potential energy wells for neutrons follow the nuclear matter densities, while those for protons have contributions from Coulomb repulsion, resulting in a bump or barrier at the nuclear edge, and wells that are less deep than the corresponding neutron wells. Nuclei with magic numbers of nucleons, corresponding to closed shells, are particularly stable. Because of pairing, nuclei with even numbers of protons and neutrons tend to be more stable than those with odd numbers. The proton and neutron energy levels are filled starting with the lowest energies and subject to the Pauli exclusion principle. Nuclear stability favors the filling of the neutron wells and the shallower proton wells to the same energy. This explains the neutron excess in medium and heavy nuclei and the tendency for isobars far from the valley floor to exhibit β-decay.

1.2.5 α decay

The α -particle is simply the nucleus of the helium atom, with mass number A = 4 and atomic number Z = 2. It consists of Z = 2 protons and A - Z = 4 - 2 = 2 neutrons. It is a very tightly bound arrangement: this ground state for two protons and two neutrons has an energy that is about 28 MeV lower than the energy of the four free nucleons. In some cases, it is energetically favorable for a nucleus of mass number A and atomic number Z to emit an α-particle, thereby producing a new nucleus, with mass number (A - 4) and atomic number (Z - 2). A case in point is the unstable isotope of uranium

$^{234}_{92}$U, containing 92 protons and 142 neutrons. It undergoes α-decay (alpha-decay) to produce an isotope of thorium, with 90 protons and 140 neutrons

$$^{234}_{92}U \rightarrow ^{230}_{90}Th + ^{4}_{2}He$$

We have seen that electric charge and mass number are conserved in α-decay process, but what of energy conservation? Well, nuclear decays will clearly involve changes in energy, much similar to atomic transitions we have seen earlier. For example, the α-decay of $^{234}_{92}$U liberates 4.77 MeV of kinetic energy, carried away (almost entirely) by the α-particle. We can write the decay as

$$^{234}_{92}U \rightarrow ^{230}_{90}Th + ^{4}_{2}He + 4.77MeV$$

and making use of Einstein's mass–energy relation, we have the mass equation

$$M(^{234}_{92}U) = M(^{230}_{90}Th) + M(^{4}_{2}He) + 4.77MeV/c^2$$

So, the liberated energy of 4.77 MeV is produced by an equivalent mass loss of $4.77MeV/c^2$. To appreciate how substantial this change in mass is, it may be compared with the mass of a proton, $m_p \simeq 1GeV/c^2$. So, in the α-decay above, the decrease in mass is about 0.50% of the proton mass, and hence about 0.002% of the mass of the original uranium nucleus, with A = 234. Quantum-mechanical tunneling is an essential mechanism for many nuclear reactions involving charged particles. α-decay occurs when α-particles inside nuclei tunnel to the outside through the Coulomb barrier. The tunneling probability depends very sensitively on the energy of the α-particle relative to the top of the barrier. This explains the huge range of α-decay lifetimes corresponding to a small range of α-particle energies. The phenomena of α–decay can be described by following theory of George Gamow, R. W. Gurney and E. U. Condon published in1982 in the framework of quantum mechanical tunnel effect. A better understanding of the process may be achieved by the semi-classical approximation known as W.K.B. approximation.

Estimation of α decay energy from semi-empirical mass formula

The α-decay can in general be represented as

$$^{A}_{Z}X \rightarrow ^{A-4}_{Z-2}Y + ^{4}_{2}He \qquad (1.10)$$

The Q-value (or the disteghration energy) of the above decay is given by

$$Q_\alpha \quad = \quad M(A, Z) - M(A - 4, Z - 2) - M(^{4}_{2}He)$$

$$
\begin{aligned}
&= B(A-4, Z-2) + B(^4_2He) - B(A, Z) \\
&= a_v(A-4) - a_s(A-4)^{2/3} - a_c\frac{(Z-2)^2}{(A-4)^{1/3}} - \frac{a_{asy}(A-2Z)^2}{(A-4)} \\
&\quad -a_vA + a_sA^{2/3} + \frac{a_cZ^2}{A^{1/3}} + \frac{a_{asy}(A-2Z)^2}{A} + B(^4_2He) \\
&= B(^4_2He) - 4a_v + a_s\left[A^{2/3} - (A-4)^{2/3}\right] \\
&\quad +a_c\left[\frac{Z^2}{A^{1/3}} - \frac{(Z-2)^2}{(A-4)^{1/3}}\right] + a_{asy}(A-2Z)^2\left[\frac{1}{A} - \frac{1}{A-4}\right] \\
&= 28.3 - 4a_v + \frac{8}{3}a_sA^{-1/3} + \frac{4a_cZ}{A^{1/3}}\left[1 - \frac{Z}{3A}\right] \\
&\quad -\frac{4a_{asy}(A-2Z)^2}{A(A-4)}
\end{aligned}
$$

$$(1.11)$$

Writing a simple FORTRAN program, Q_α can be computed easily substituting the values of the coefficicients in MeV unit. Binding energy of ^{4}He is taken as 28.3Mev in the above formula. One can verify that $Q_\alpha > 0$ for A>160 meaning the those nuclei beyond mass 160 should be α unstable. But in practice α decay is observed in nuclei in the mass region A>200. For 160<A<200, the energy released is small enough to allow for barrier pentration.

1.2.6 β-decay and electron capture

The usual type of β-decay (beta-decay) involves the emission of an electron from the nucleus of an atom. The process occurs when a neutron in the original nucleus transforms into a proton, so increasing the atomic number by one. For reasons that are not important here, another particle is created in thebeta-decay process too. It is called the electron anti-neutrino and it has zero electric charge. Creation of an electron and an electron anti-neutrino occurs in what is called β^--decay (beta-minus decay), the minus sign indicating that the electron is negatively charged. A nucleus that undergoes β^--decay is the unstable lead isotope $^{214}_{82}$Pb which transforms into a stable bismuth isotope $^{214}_{83}$Bi.The decay in this case can be represented as

$$^{214}_{82}Pb \rightarrow ^{214}_{83}Bi + e^- + \overline{\nu_e}$$

The rather clumsy symbol $\overline{\nu_e}$ represents an electron anti-neutrino which has zero charge. The subscript e indicates that it is associated with an electron, and the bar over the top of the letter indicates that it is an antiparticle.

The process described above is only half of the story as far as beta-decay is concerned. There is a very closely related process, called β^+-decay (beta-plus decay or sometimes inverse beta decay), in which a positively charged particle, called a positron, is created, along with an electron neutrino, which has zero charge. In this process, a proton in the original nucleus transforms into a neutron, so decreasing the atomic number by one. A nucleus that undergoes β^+-decay is the unstable oxygen isotope $^{14}_8$O which transforms into a stable nitrogen isotope$^{14}_7$N. The decay in this case can be represented as

$$^{14}O \rightarrow^{14}_7 N + e^+ + \nu_e \tag{1.12}$$

Here, the symbol e^+ is used to represent the positron (also known as an anti-electron) and ν_e is the electron neutrino. The observable consequence of β^+-decay is that the positron produced will immediately combine with an electron to produce gamma-rays:

$$e^+ + e^- \rightarrow \gamma \, [E_\gamma = 1.022 MeV] \tag{1.13}$$

Related to β^+-decay is the process of electron capture, in fact the outcome is virtually the same. Some nuclei which have too many protons, rather than undergoing β^+-decay, instead capture an electron into the nucleus. As a result, a proton is transformed into a neutron and an electron neutrino is emitted. A nucleus that undergoes electron capture is the unstable beryllium isotope $^7_4$Be which transforms into a stable lithium isotope $^7_3$Li. The decay in this case can be represented as

$$^7_4 Be + e^- \rightarrow^7_3 Li + \nu_e \tag{1.14}$$

As in β^+-decay, the atomic number decreases by one and the mass number remains the same. The condition for different modes of beta-decay follows:
β^- : M (A, Z) > M (A, Z+1)
β^+: M (A, Z) > M (A, Z-1) $+2m_e$ $[m_e = 0.511 MeV]$
e-capture: M (A, Z)>M (A, Z-1)

Estimation of β decay energy from semi-empirical mass formula

The β-decay process can in general be represented as

$$\beta^+ : {}^A_Z X \rightarrow^A_{Z-1} Y + \beta^+ + \nu_e \tag{1.15}$$

$$\beta^- : {}^A_Z X \rightarrow^A_{Z+1} Y + \beta^- + \overline{\nu_e} \tag{1.16}$$

As evident from the above relations, in β decay process the parent and daughter nuclei are isobar and they are also mirror pair having a one unit difference between proton and neutron numbers. Thus for mirror pairs Z-N=1, which gives A=2Z-1. In mirror pair, nuclus of higher Z are naturally found to be β^+ emitter. Example in the pair $(^{13}N, ^{13}C)$, ^{13}N is the β^+ emitter. From Semi-empirical mass formula for odd-A nuclei (for which $\delta = 0$), we have

$$
\begin{aligned}
M(A, Z) &= ZM_H + NM_n - a_v A + a_s A^{2/3} \\
&\quad + \frac{a_c Z^2}{A^{1/3}} + \frac{a_{asy}(A - 2Z)^2}{A} \\
&= ZM_H + (Z-1)M_n - a_v A + a_s A^{2/3} \\
&\quad + \frac{a_c Z^2}{A^{1/3}} + \frac{a_{asy}}{A} \quad (1.17) \\
M(A, Z-1) &= (Z-1)M_H + ZM_n - a_v A + a_s A^{2/3} \\
&\quad + \frac{a_c(Z-1)^2}{A^{1/3}} + \frac{a_{asy}}{A} \quad (1.18) \\
M(A, Z) &- M(A, Z-1) = M_H - M_n + a_c \frac{(2Z-1)}{A^{1/3}} \quad (1.19) \\
Q_{\beta^+} &= M(A, Z) - M(A, Z-1) - 2m_e \\
&= M_H - M_n + a_c \frac{(2Z-1)}{A^{1/3}} - 2m_e \\
&= a_c A^{2/3} - (M_n - M_H + 2m_e) \ [since, A = 2Z - 1] \\
&= a_c A^{2/3} + 1.804 MeV \quad (1.20)
\end{aligned}
$$

where $a_c = \frac{3}{5}(\frac{e^2}{4\pi\epsilon_0 r_0}) = 0.7104 MeV$. In the simlar way β^- disintegration energy cal also be estimaed from

$$
Q_{\beta^-} = M(A, Z) - M(A, Z+1) \quad (1.21)
$$

1.2.7 γ-decay

The final type of nuclear decay that we consider here is γ-decay (gamma-decay). In contrast to the two processes of α-decay and β-decay, this involves no change in the numbers of neutrons and protons. Gamma-decay occurs when a nucleus finds itself in an excited state. A quantum jump down to the ground-state configuration of the same number of neutrons and protons is accompanied by the emission of a photon, as with electron transitions in atoms. This time however, the photon energy is around a million time larger– it is a gamma-ray photon. Such excited states of nuclei may be created in the process of α-decay or β-decay, or by the collisions.[2]

A nucleus initially in a higher excited state (N*) having angualar mmonetum J_i, and parity π_i may make transition to some lower excited (or ground) state (N) of angular momentum and parity J_f, π_f respectively obeying the selection rule:

i) $\pi_i \pi_f = (-1)^\lambda$, for E$\lambda$-type transition and

ii)$\pi_i \pi_f = (-1)^{\lambda+1}$, for M$\lambda$-type transition

where $|J_i - J_f| \leq \lambda \leq (J_i + J_f)$

1.2.8   Nuclear fusion

The most important nuclear process that is responsible for the energy production in stars and has great role in astrophysics is nuclear fusion. Many nuclear reactions between charged particles occur by the particles tunneling through their mutual Coulomb barrier so that they can interact by the strong nuclear force. Fusion reactions in the Sun can only occur at the prevailing temperatures by tunneling. As noted above, certain configurations of nucleons are energetically more favorable than others. Hence, by forcing certain nuclei sufficiently close together (overcoming their mutual electrical repulsion); they may fuse to form a single more massive nucleus which is at a lower energy than the original two nuclei. Once again, an example should make the process clear. A nuclear fusion process which occurs in the majority of stars is the fusion of two nuclei of helium-3. It may be represented as

$$\tfrac{3}{2}He +\tfrac{3}{2} He \to\tfrac{4}{2} He + p + p + 12.86 \, MeV \tag{1.22}$$

The mass of each nucleus of helium-3 ($\tfrac{3}{2}$He) is $2808 MeV/c^2$, whilst the mass of a nucleus of helium-4 ($\tfrac{4}{2}$He) is $3727 MeV/c^2$ and the mass of a proton is $938 MeV/c^2$. Simply adding up these masses, there is a deficit of $\sim 13 MeV/c^2$ on the right-hand side. This lost mass appears as Q-energy, liberated by the fusion process. Examples of some other nuclear fusion reactions are

$$\tfrac{2}{1}H +\tfrac{2}{1} H \quad \to \quad \tfrac{3}{2}He + n + 3.26 MeV \tag{1.23}$$

$$\tfrac{2}{1}H +\tfrac{2}{1} H \quad \to \quad \tfrac{3}{1}H + p + 4.04 MeV \tag{1.24}$$

$$\tfrac{3}{1}H +\tfrac{2}{1} H \quad \to \quad \tfrac{4}{2}He + n + 17.6 MeV \tag{1.25}$$

---

[2]A collision is an event in which two or more bodies exert forces on each other for a relatively short time. Specifically, collisions can either be elastic, meaning they conserve both momentum and kinetic energy, or inelastic, meaning they conserve momentum but not kinetic energy. Inelastic collision plays vital role in nuclear reaction dynamics.

$$\begin{align}
{}^3_2He + {}^2_1H &\rightarrow {}^4_2He + p + 18.3 MeV \tag{1.26}\\
{}^6_3Li + {}^2_1H &\rightarrow {}^4_2He + {}^4_2He + 22.4 MeV \tag{1.27}\\
{}^7_3Li + {}^1_1H &\rightarrow {}^4_2He + {}^4_2He + 17.3 MeV \tag{1.28}
\end{align}$$

The above nuclear fusion reaction can occur only if the interacting particles have sufficient kinetic energy to reach each other overcoming the electrostatic repulsion. For two deuteron, to undergo fusion reaction they must come closer to within

$$R = r_0(A_1^{1/3} + A_2^{1/3}) \approx 3.02 \times 10^{-15}m \ [r_0 = 1.2 \times 10^{-15}m] \tag{1.29}$$

where R is of the order of separation between the centers of the two approaching deuteron nuclei for which the strength of the electrostatic potential will be

$$V = \frac{e^2}{8\pi\epsilon_0 R} = \frac{\alpha c \hbar}{2R} \simeq 0.24 \ MeV \tag{1.30}$$

To initiate such reactions, high temperatures (of the order $10^7 K$ or higher) are required to provide the necessary thermal kinetic energy to overcome the electrical repulsion (or the Coulomb barrier) between interacting nuclei as all of them are positively charged. As evident from Eq. (1.19),the height of the Coulomb barrier depends on the charges and radii of the interacting nuclei. In thermo-nuclear reaction studies, it is convenient to express temperature in terms of the kinetic energy of the particles involved in the interaction via the relation $K = K_B T$, K being the kinetic energy corresponding to the most probable speed of the interacting particles at the absolute temperature T Kelvin and K_B is the Boltzmann constant. Thus the temperature at the centre of the sun be better said 1.3keV rather than saying $1.5 \times 10^7 K$. Height of the Coulomb barrier for two interacting protons is nearly equal to 400keV which is much higher than the temperature (1.3keV) at the core of the sun. Thus it seems that thermonuclear fusion cannot occur in the interior of the sun but we know that this is the dominant feature in the sun as well as in more massive stars. In practice the thermonuclear fusion reaction occurs by the quantum mechanical tunneling of the interacting particles (like protons, deuterons etc.) through their mutual Coulomb barriers. Following nuclear stability curve (i.e., or the curve obtained by plotting average binding energy per nucleon (B/A) against mass number A), it can be said that in fusion of light nuclei or when various light nuclei are combined together energy will be liberated. So, in the core of stars, nuclear fusion processes liberate energy to power the star and also convert light elements into heavier ones. The limiting mass at which the process ceases to be energetically favorable is found to lie around that of the nuclei of iron, cobalt and nickel. So, nuclear fusion provides an energy source for stars only until their cores are composed of nuclei such as iron. For many stars, nuclear fusion will end long before iron is formed as the temperatures in their cores are not sufficient to trigger fusion reactions beyond, say, helium or carbon. The life cycles of stars are closely dependent on these energy dependent processes. As nuclear fusion is of utmost importance in nuclear astrophysics we will elaborate it little bit in context of Coulomb barrier penetration probability.

1.2.9   The Coulomb barrier

The physical difficulty in achieving fusion reactions is due to the fact that, unlike neutron absorption in fission reactions, the nuclei which interact are charged and, therefore, have a Coulomb repulsion. In order for the nuclei to have strong interactions, they must approach one another at a distance of the order of the range of nuclear forces, i.e. the radius of the nuclei. This is once again a situation where one must cross an electrostatic potential barrier. The barrier has a height given by

$$\frac{Z_1 Z_2 e^2}{4\pi\epsilon_0 a} = \frac{Z_1 Z_2 \alpha \hbar c}{a} = \frac{1.4 Z_1 Z_2}{a} MeV fm, \tag{1.31}$$

where a is the distance within which the attractive nuclear forces become larger than the Coulomb force. For energies (or temperatures) less than $\sim$ 1 MeV the barrier is operative. If E is the energy of the particle impinging on this barrier, the probability to cross the barrier by quantum tunneling is proportional to the Gamow factor (G) defined as

$$P \propto G = \exp\left[-2\int_a^b dr \sqrt{\frac{2\mu}{\hbar^2}(V(r) - E)}\right] \tag{1.32}$$

where $\mu = \frac{m_1 m_2}{m_1 + m_2}$ is the reduced mass of the two interacting nuclei and b is the classical turning point defined by V (b) = E, where V (r) is the repulsive Coulomb potential. The integral on the right hand side of (1.20) can easily be calculated, if

$$a \ll b = E_B = \frac{Z_1 Z_2 e^2}{4\pi\epsilon_0 E} = \frac{Z_1 Z_2 \alpha \hbar c}{E} = \frac{1.43 Z_1 Z_2}{E} MeV fm, \tag{1.33}$$

and therefore the radius a ($\approx r_0(A_1^{1/3} + A_2^{1/3})$, $r_0 = 1.2 fm$) can be taken equal to zero in a good approximation leading to the following expression

Table 1.2: Some fusion reactions. The first three are used in terrestrial fusion reactors. The last three make up the "PPI" cycle responsible for most of the energy generation in the Sun. Note the tiny S(E) for the weak-reaction $pp \to de^+\nu_e$. It can only be calculated using weak-interaction theory.

Reaction	Q (MeV)	S(10keV) (keV b)	E_B (keV)	E_G(1keV) (keV)	E_G(20keV) (keV)
$dd \to n^3He$	3.25	58.3	987.0	5.1	37.5
$dd \to p^3H$	4.0	57.3	987.0	5.1	37.5
$dt \to n^4H$	17.5	14000.0	1185.0	6.8	50.1
$pp \to de^+\nu_e$	1.442	3.8×10^{-22}	526.0	5.1	37.5
$pd \to^3 He\gamma$	5.493	2.5×10^{-4}	701.0	5.6	41.2
$2^3He \to pp^4He\gamma$	12.859	5×10^3	25200.0	18.5	136.0

due to Gamow (in 1934) for the tunneling probability:

$$P \sim \exp\left(-\frac{2\pi Z_1 Z_2 e^2}{4\pi\epsilon_0 \hbar v}\right) = \exp(\sqrt{-E_B/E}) \tag{1.34}$$

where $v = \sqrt{\frac{2E}{\mu}}$ is the relative velocity for the center-of-mass kinetic energy E. The barrier is characterized by the parameter

$$E_B = 2\pi^2 Z_1^2 Z_2^2 \alpha^2 \mu c^2 = 1.052 \times Z_1^2 Z_2^2 \mu c^2 /(keV) \tag{1.35}$$

We note that the argument of the exponential increases in absolute value with the product of the charges and that it decreases as the inverse of the velocity. The higher the energy of the nuclei is, the greater the probability to tunnel through the barrier. Likewise, the larger the product of charges $Z_1 Z_2$ is, the higher the barrier, therefore at a given energy, it is the lighter nuclei which can undergo fusion reactions. For particles of charge +1 (e.g. d + d) we have

$$E = 1keV \Rightarrow P \sim 10^{-13}$$

$$E = 10keV \Rightarrow P \sim 10^{-3}$$

This suggests that $\sim 10keV$ is the order of magnitude of the kinetic energy that the nuclei must have in order for fusion reactions to take place. (Much below that energy, the cross section vanishes for all practical purposes.) The energy of the nuclei comes from their thermal motion, therefore from the temperature of the medium where they are contained. Hence the name thermonuclear reactions for fusion reactions. We must find the factors of proportionality between the tunneling probability and the reaction cross section. For absorption reaction inhibited by potential barriers, we the cross-section should be of the form

$$\sigma(E) = \frac{S(E)}{E} \exp\left(-\sqrt{E_B/E}\right) \tag{1.36}$$

where S(E) is a slowly varying function of the center-of-mass energy and E_B is given by (1.23). The experimental determination of the nuclear factors S(E) is a problem of major interest for all calculations in astrophysical and cosmological nucleosynthesis, as we shall see later. Examples are shown in Figures 1.5 and 1.6. Considerable effort are being made in recent years to measure the cross-sections at energies comparable to stellar temperatures. Without such data it was necessary to extrapolate the S(E). Table 1.2 lists the S(E) for some important fusion reactions. It is to be noted that the tiny cross-section for the stellar reaction $pp \rightarrow {}^2H + e^+ + \nu_e$ makes this reaction unobservable. The S(E) must therefore be calculated using weak-interaction theory. The Gamow formula (1.25) can also be derived easily using Born approximation. A cross-section involves, like any transition rate, the square of a matrix element between initial and final states $| < \psi_f|M_{if}|\psi_i|^2$. When forces are short range (like the strong nuclear force), the asymptotic states $|\psi_i >$ and $|\psi_f >$ are monochromatic plane waves. In the presence of a Coulomb interaction, which is of infinite range, the asymptotic behavior is different. The wave function is exactly calculable[3] and, asymptotically, the argument of the exponential has additional terms of the form

$$\zeta(\mathbf{r}) \sim \exp(i(\mathbf{k.r} + \gamma \log kr) \tag{1.37}$$

with

$$\gamma = \frac{Z_1 Z_2 e^2}{4\pi\epsilon_0 \hbar v} \tag{1.38}$$

These asymptotic wave functions are called Coulomb scattering states. Consider, for instance, the reaction $d + t \rightarrow^4 He + n$. In the initial state, we must use Coulomb scattering states (and not the usual asymptotic states). Since the strong interaction is short range, we can, in good approximation, simply multiply the usual matrix element by the value of the Coulomb scattering wave at the origin

$$\phi_{Coul}(0) = \Gamma(1 + i\gamma)e^{-\pi\gamma/2} \tag{1.39}$$

---

[3]A. Messiah, Quantum Mechanics vol. 1, chap. XI-7

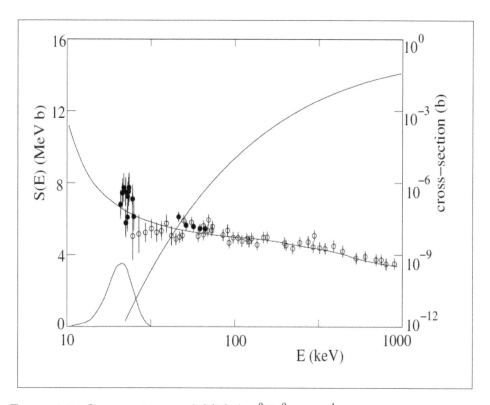

Figure 1.5: Cross-section and S(E) for ^{3}He^{3}He $\rightarrow$ ^{4}He pp, as measured by the LUNA underground accelerator facility. The top panel shows the small ($\sim 1m^2$) experiment consisting of a ^{3}He ion source, a 50 kV electrostatic accelerator, an analyzing magnetic spectrometer, a gaseous (^{3}He) target chamber, and a beam calorimeter to measure the beam intensity. The sides of the target chamber are instrumented with silicon ionization counters that measure dE/dx and E of protons produced by $^3He^3He \rightarrow^4 Hepp$ in the chamber. Because of the very small cross-sections to be measured, the experiment is in the deep underground laboratory LNGS, Gran Sasso, Italy, where cosmic-ray background is eliminated. The bottom panel shows the LUNA measurements as well as higher energy measurements. The lowest energy measurements cover the region of the solar Gamow peak for this reaction (Figure 1.7). Note that while the cross section varies by more than 10 orders of magnitude between E = 20 keV and 1MeV, the factor S(E) varies only by a factor ~ 2.. [Ref: M. Junker et al., Phys. Rev. C 57 (1998) 2700]

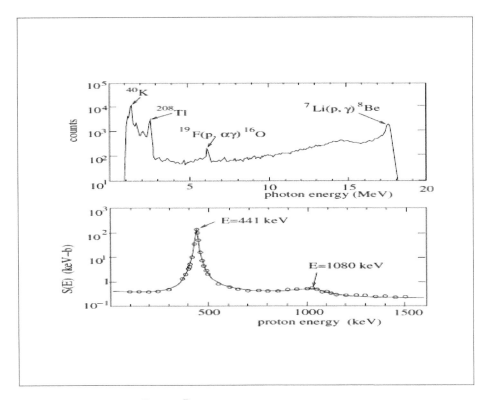

Figure 1.6: S(E) for $p^7Li \to^B e\gamma$ reaction. The top panel shows how a proton beam impinges upon a target consisting of $10\mu gcm^{-2}$ of LiF evaporated on a copper backing. The target is inside a large NaI scintillator that detects photons emerging from the target. The middle panel shows a typical photon energy spectrum showing peaks due to $^7Li(p,\gamma)^8Be$, in addition to peaks due to $^{19}F(p,\alpha\gamma)^{16}O$ and to natural radioactivity in the laboratory walls. The S-factor deduced from the photon counting rate is shown on the bottom panel as a function of proton energy. It shows the presence of two resonances due to excited states of ^8Be.[Ref: D. Zahnow et al., Z. Phys. A 351 (1995) 229-236]

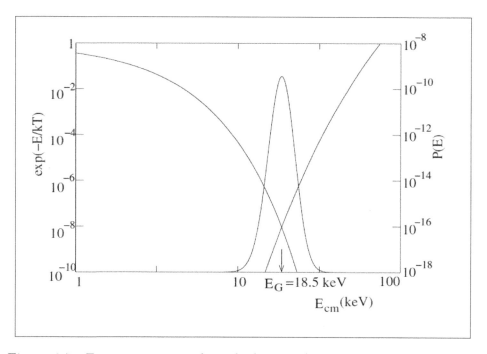

Figure 1.7: Factors entering the calculation of the pair reaction rate Eq. (1.37). The Boltzmann factor $\exp(-E/kT)$ (logarithmic scale on the left) and the barrier penetration probability $P(E) = \exp(-\sqrt{E_B/E})$ Eq. (1.23) (logarithmic scale on the right) are calculated for kT = 1 keV (corresponding to the center of the Sun) and for the reaction $^3He\,^3He \rightarrow^4 Hepp$. The product is the Gaussian-like curve in the center (shown on a linear scale). It is maximized at $E_G = (\sqrt{E_B}kT/2)^{2/3} \sim 18.5keV$ and most reactions occur within $\sim 5keV$ of this value. Note the small values of $\exp(-E_{G_m}/kT) \sim 10^{-8}$ and $P(E_G) \sim 10^{16}$.

i.e. multiply the nuclear cross-section by the factor

$$|\Gamma(1+i\gamma)e^{-\pi\gamma/2}|^2 = \frac{\pi\gamma e^{-\pi\gamma}}{\sinh\pi\gamma} \qquad (1.40)$$

Once the kinematic factors are taken into account, one recovers the tunnel effect factor

$$e^{-2\pi\gamma} \sim \frac{e^{-\pi\gamma}}{\sinh\pi\gamma} \qquad (1.41)$$

introduced empirically for $\gamma \gg 1$.

1.2.10   Reaction rate in a medium

For the specific case of a mixture of deuterium and tritium, the reaction rate per unit volume which is the number (say, R) of reactions per unit volume per unit time can be determined. If number densities of d and t be n_d and n_t respectively and the deuterium nucleus has a velocity v with respect to the tritium nucleus, then the probability per unit time λ that a fusion reaction occurs is:

$$\lambda = n_t\sigma(v)v \qquad (1.42)$$

where σ is the fusion cross-section of (1.25). The rate per unit volume is then found by multiplying the rate per deuterium nucleus by n_d :

$$R = n_d n_t \sigma(v)v \qquad (1.43)$$

We must average this expression over the velocity distribution in the medium at temperature T:

$$R = n_d n_t < \sigma(v)v > \qquad (1.44)$$

where $< \sigma(v)v >$ is the average of the product $\sigma(v)v$, the probability of v being determined by the Maxwell distribution at temperature T. Because of the decreasing Coulomb barrier, the product $\sigma(v)v$ increases rapidly with the energy. In the averaging, it is however in competition with the decrease of the Maxwell distribution with increasing velocity. If only one of the two species is in motion, we would have

$$< \sigma v > \sim \int d^3v e^{-E/kT}\sigma(v) \sim \int v^3 e^{-mv^2/2kT}\sigma(v)dv \qquad (1.45)$$

In reality, both species are in motion so the integral is slightly more complicated. For nuclear of masses of m_1 and m_2 , we have

$$< \sigma v > = \left(\frac{m_1 m_2}{(2\pi kT)^2} \right)^{3/2} \int d^3 v_1 d^3 v_2 e^{-(m_1 v_1^2 + m_2 v_2^2)/2kT} \sigma(v) v \qquad (1.46)$$

where $v = |\vec{v_1} - \vec{v_2}|$ is the relative velocity. Turning to center-of-mass variables, $\mu = \left(\frac{m_1 m_2}{m_1 + m_2} \right)$ being the reduced mass, we can integrate over the total momentum (or the velocity of the center of gravity). This leads to

$$< \sigma v > = \sqrt{\frac{8}{\pi \mu (kT)^3}} \int e^{-E/kT} E \sigma(E) dE \qquad (1.47)$$

Using (1.25) this becomes

$$< \sigma v > = \sqrt{\frac{8}{\pi \mu (kT)^3}} \int dE e^{-\sqrt{E_B/E}} e^{-E/kT} S(E) \qquad (1.48)$$

The integrand contains the product of two exponentials shown in Figure 1.7. The exponential term $\exp(-\sqrt{E_B/E})$ is smaller at smaller energy while the other term $\exp(-\frac{E}{kT})$ is smaller at larger energy. Let the the product peaks at some energy E_{max} such that

$$\frac{d}{dE} \left(\sqrt{E_B/E} + E/kT \right) = 0$$

at $E = E_{max} = (\frac{1}{2} kT E_B^{1/2})^{2/3}$. This E_{max} is called the effective burning energy or the Gamow energy

$$E_G \; = \; E_{max} = E_B^{1/3} (kT/2)^{2/3}, \qquad (1.49)$$

where E_B is given by Eq. (1.24). As long as S(E) has no resonances (e.g. as in Figure 1.6) only the narrow region around the indexGamow energy (called the Gamow peak) contributes significantly to $\sigma(v)v$. Its position determines the effective energy at which the reaction takes place. In the absence of resonances, the nuclear factor S(E) varies slowly and only the value $S(E_G)$ is relevant so it can be taken out of the integral (1.37). We can also make a Taylor expansion of the argument of the exponential in the region E_G:

$$\sqrt{E_B/E} + E/kT \sim 3/2 \left(\frac{2E_B}{kT} \right)^{1/3} + \frac{1}{2} \frac{(E - E_G)^2}{\Delta_E^2} \qquad (1.50)$$

where the width of the Gamow peak is

$$
\begin{aligned}
\Delta_E &= \frac{2}{\sqrt{3}} E_G \left(\frac{kT}{E_G} \right)^{1/2} \\
&= \frac{2}{\sqrt{3}} E_B^{1/6} \left(\frac{kT}{2} \right)^{5/6}
\end{aligned}
\tag{1.51}
$$

It is to note that the Gamow peak is relatively narrow:

$$
\Delta E / E_G \sim (kT/E_B)^{1/6}
\tag{1.52}
$$

We then have

$$
\begin{aligned}
< \sigma v > &= \frac{8\pi}{\sqrt{\mu}} (kT)^{-3/2} S(E_G) \exp \left[-(3/2) \left(-\frac{2E_B}{kT} \right)^{1/3} \right] \\
&\times \int \exp \left(\frac{(E - E_G)^2}{2\Delta_E^2} \right) dE
\end{aligned}
\tag{1.53}
$$

The Gaussian integral just gives a factor ΔE so we end up with

$$
< \sigma v >= \frac{8\pi}{\sqrt{\mu}} (kT)^{-2/3} E_B^{1/6} S(E_G) \exp \left[-(3/2) \left(-\frac{2E_B}{kT} \right)^{1/3} \right]
\tag{1.54}
$$

 Figure 1.8 shows σv as a function of temperature for $dd \rightarrow n^3\text{He}$ and for $dt \rightarrow n^4\text{He}$. The rate rises rapidly for $kT < 10 keV$ before leveling off. We can say that $kT \sim 10 keV$, i.e. $T \sim 1.5 \times 10^8 K$, defines an optimal temperature for a fusion reactor.

1.2.11   Compound nuclear hypothesis

The compound nuclear hypothesis, propounded by Niels Bohr, is based on his observations on nuclear scattering and reaction experiments. According to the hypothesis when a nuclear projectile x enters a nuclear target X to initiate a nuclear reaction, an intermediate nucleus (C*) is formed before the production of the final product nuclei y and Y or the emission of γ-rays. This intermediate nucleus having relatively longer lifetime ($\sim 10^{-15}s$) is called the compound nucleus and the process can be represented by

$$
X + x \rightarrow C^* \rightarrow Y + y
$$

or

$$
X(x, y)Y
$$

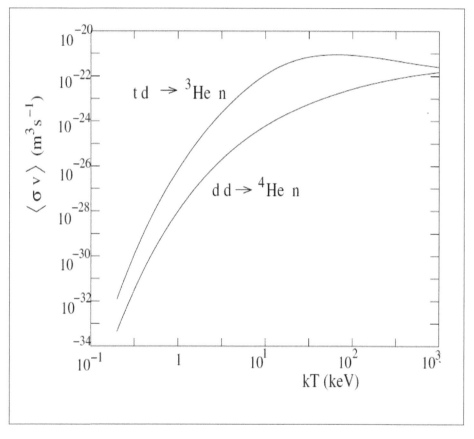

Figure 1.8: Variation of the pair reaction rate $< v\sigma >$ as a function of the temperature for d-d and d-t mixtures.

The incoming projectile x on entering the target nucleus X quickly dissipates its energy and merges with the closely packed nucleons thereby disturbing the random motion of the nucleons inside the nucleus. In the process none of the single nucleon is able to acquire sufficient energy from the projectile to get emitted from the nucleus. However after a long time when a very large number of collisions($\sim 10^7$) among the nucleons have taken place, one of the nucleons may accumulate sufficient energy to escape from the nucleus thereby leading the residual nucleus to de-excite (cool off) to the ground state. This phenomenon can be compared to the evaporation mechanism of a heated liquid drop containing large number of molecules. The mean time interval between two successive collisions is about ($R/v = 2 \times 10^{-15}/5 \times 10^7$) $10^{-22}s$. So the lifetime of the compound nucleus is of the order of $10^7 \times 10^{-22}s \sim 10^{-15}s$.

Compound nucleus being relatively longer lived, nuclear reaction proceeds in two stages: 1) formation of the CN by the absorption of the projectile (x) by the target (X), and 2) disintegration of CN in the reaction products y and Y, in a manner independent of its formation history (known as Bohr's independence hypothesis). Sometimes the residual nucleus (Y) is left in a highly excited state which "boils off" another particle y' leading to a two particle emission process $X(x, yy')Y'$ and the process may continue further with the emission of y'' and leaving excited Y''. The two stages of reaction processes can be represented symbolically as

$$X + x \rightarrow C^* \quad \rightarrow \quad Y + y$$
$$\rightarrow \quad C + \gamma$$

The probability decay (measured in terms of the level width, Γ) being reciprocal of the mean-life τ of the compound nucleus, we have from uncertainty relation

$$\Gamma \sim \hbar/\tau$$

Now if Γ_y represent the partial level width for decay with the emission of y, then considering various particle emissions in the reaction we may get the total width of levels as

$$\Gamma \quad = \quad \sum_y \Gamma_y + \Gamma_\gamma = (\Gamma_y + \gamma_{y'} + \Gamma_{y''} + ...) + \Gamma_\gamma \qquad (1.55)$$

and the relative probabilities of different types of decay as

$$\eta_y \quad = \quad \frac{\Gamma_y}{\Gamma}, \eta_{y'} = \frac{\Gamma_{y'}}{\Gamma}, \eta_{y''} = \frac{\Gamma_{y''}}{\Gamma}, ..., \eta_\gamma = \frac{\Gamma_\gamma}{\Gamma} \qquad (1.56)$$

Because of the independence hypothesis of CN decay, we may write the cross-section for the reaction process X(x,y)Y as a product of the cross-section σ_x for the formation of the compound nucleus and the probability of its decay

$$\sigma(x, y) \quad = \quad \sigma_x \eta_y = \sigma_x \frac{\Gamma_y}{\Gamma} \tag{1.57}$$

The above relation is written considering only one particular energy state of C^* or only one resonance. This is possible if the energy levels are well separated and are so sharp that they do not interfere themselves.

1.2.12   Breit-Wigners one-level formula

The state of compound nucleus can be represented by a damped harmonic wave

$$\psi(t) \quad = \quad \psi_0 e^{-iE_r t/\hbar} e^{-\Gamma t/2\hbar} \tag{1.58}$$
$$= \quad \psi_0 e^{-i(E_r - i\Gamma/2)t/\hbar} \tag{1.59}$$

where $\Gamma/2$ is the half-width of the decaying level of life-time $\tau = \hbar/\Gamma$. Although the above function does not represent a stationary state, it may be assumed to be built up by the superposition of stationary states of different energies by Fourier integral method:

$$\psi(t) = \int_{-\infty}^{\infty} A_E e^{-iEt/\hbar} dE \tag{1.60}$$

By FT we have

$$A_E \quad = \quad \frac{1}{2\pi} \int_0^\infty \psi(t') e^{iEt'/\hbar} dt'$$
$$= \quad \frac{1}{2\pi} \int_0^\infty \psi_0 e^{i(E - E_r + i\Gamma/2)t'/\hbar} dt'$$
$$= \quad \frac{\psi_0}{2\pi} \left[\frac{e^{i(E - E_r + i\Gamma/2)t'/\hbar}}{i(E - E_r + i\Gamma/2)t'/\hbar} \right]_0^\infty$$

As the upper limit term vanishes due to damping term $e^{-\Gamma t/2\hbar}$, we get

$$A_E \quad = \quad \frac{\psi_0}{2\pi} \frac{i\hbar}{E - E_r + i\Gamma/2} \tag{1.61}$$
$$\mid A_E \mid^2 \quad = \quad \frac{\mid \psi_0 \mid^2}{4\pi^2} \frac{\hbar^2}{(E - E_r)^2 + \Gamma^2/4} \tag{1.62}$$

But, the cross-section for the formation of compound nuclear state E_C in the process X+x is proportional to the squared amplitude. Hence, we may write

$$\sigma_x = \frac{C}{(E - E_r)^2 + \Gamma^2/4} \tag{1.63}$$

where C is a constant. Applying reciprocity theorem the proportionality constant C is found to be

$$C = \frac{\pi}{k^2}\Gamma_x\Gamma \tag{1.64}$$

Using Eqs. (1.52) and (1.53) in Eq. (1.46) we get the cross-section for the reaction X(x,y)Y

$$\sigma(x,y) = \frac{\pi}{k^2}\frac{\Gamma_x\Gamma_y}{(E - E_r)^2 + \Gamma^2/4} \tag{1.65}$$

The above Eq. (1.54) is known as Breit-Wigner one level formula for spin less nuclei at very low energies for which the relative angular momentum (l) of the particles in the entrance channel is zero i.e., l=0.

1.2.13   Cross-sections and their derivations

In nuclear physics we often encounter with scattering and reaction cross-sections as well as with the terms differential and total cross-sections. The number of particles scattered into unit solid angle per unit time, per unit incident flux, per target point, is called differential scattering cross-section. It is defined as

$$\frac{d\sigma}{d\Omega} = \mid f(\theta,\phi) \mid^2$$

If the incident plane wave e^{ikz} is so normalized that there is only one particle per unit volume, then

$$n_{inc} = \mid \psi_{inc} \mid^2 = \mid e^{ikz} \mid^2 = 1$$

So, the incident flux which is the number of particles incident per unit area is

$$v \mid \psi_{inc} \mid^2 = v,$$

v being the velocity of the incident particles. The probability of scattering by a single nuclear target into the solid angle $d\Omega$ is

$$\frac{d\sigma}{d\Omega}v \mid \psi_{inc} \mid^2 d\Omega = \frac{d\sigma}{d\Omega}vd\Omega$$

Now, the number of scattered particles per unit volume in direction (θ, ϕ) is $\mid \psi_{sc} \mid^2 = \mid f(\theta, \phi) \mid^2 /r^2$, so the scattered flux will be

$$v \mid \psi_{sc} \mid^2 = v \mid f(\theta, \phi) \mid^2 /r^2$$

Thus, the number of particles scattered into the solid angle $d\Omega$ and passing through an elemental area dS at a distance r from the scattering centre is

$$(scattered\ flux) \times dS = v \mid f(\theta, \phi) \mid^2 /r^2.r^2 d\Omega = v \mid f(\theta, \phi) \mid^2 d\Omega.$$

Comparing the preceding relations we have the differential scattering cross-section

$$\frac{d\sigma}{d\Omega} = \mid f(\theta, \phi) \mid^2$$

[Note: In QM, flux means the probability current density given by

$$J = \frac{i\hbar}{2m}[\psi\nabla\psi^* - \psi^*\nabla\psi]$$

For incident wave,

$$\psi_{inc} = e^{ik.r}$$

,

$$J_{inc} = \frac{i\hbar}{2m}[e^{ik.r}\psi\nabla_r e^{-ik.r} - e^{-ik.r}\nabla_r e^{ik.r}] = \frac{\hbar k}{m}]$$

The total scattering cross-section is defined as

$$\sigma_s = \int_\Omega \frac{d\sigma}{d\Omega}d\Omega = \int_\Omega \mid f(\theta, \phi) \mid^2 sin\theta d\theta d\phi$$

Derivation of scattering cross-section (σ_{sc}) formula

A beam of mono-energetic incident particles proceeding along Z-direction can be represented by a plane wave

$$\psi_{inc} = e^{ikz} = e^{ikr cos\theta} = \sum_{l=0}^{\infty} i^l(2l+1)j_l(kr)P_l(cos\theta) \tag{1.66}$$

which in the asymptotic limit (ie. r$\rightarrow \infty$) can be expanded as

$$\psi_{inc} = \sum_{l=0}^{\infty} i^l(2l+1)\frac{sin(kr - l\pi/2)}{kr}P_l(cos\theta)$$

$$= \frac{1}{2ikr}\sum_{l=0}^{\infty} i^l(2l+1)[e^{i(kr-l\pi/2)} - e^{-i(kr-l\pi/2)}]P_l(cos\theta) \tag{1.67}$$

Thus the incident plane wave can be represented as superposition of a set of outgoing spherical wave e^{ikr}/r and a set of incoming spherical wave e^{-ikr}/r of equal amplitude. In the presence of the scattering nucleus, only the outgoing spherical wave will be affected in amplitude as well as in phase, and the total wave function can be expressed as

$$\psi(r,\theta) = \frac{1}{2ikr} \sum_{l=0}^{\infty} i^l (2l+1)[\eta_l e^{i(kr-l\pi/2)} - e^{-i(kr-l\pi/2)}]P_l(cos\theta) \quad (1.68)$$

Again, since the total wave function ψ in the presence of scatterer can also be represented as

$$\psi = \psi_{inc} + f(\theta,\phi)e^{ikr}/r \qquad (1.69)$$

we have,

$$
\begin{aligned}
f(\theta,\phi) &= re^{-ikr}[\psi - \psi_{inc}] \\
&= (re^{-ikr})\frac{1}{2ikr}\sum_{l=0}^{\infty} i^l(2l+1) \\
&\times [\eta_l e^{i(kr-l\pi/2)} - e^{-i(kr-l\pi/2)} - e^{i(kr-l\pi/2)} + e^{-i(kr-l\pi/2)}]P_l(cos\theta) \\
&= \frac{1}{2ik}\sum_{l=0}^{\infty} i^l(2l+1)[(\eta_l-1)e^{-il\pi/2)}]P_l(cos\theta) \\
or, \ f(\theta,\phi) &= \frac{1}{2ik}\sum_{l=0}^{\infty}(2l+1)[(\eta_l-1)]P_l(cos\theta) \qquad (1.70)
\end{aligned}
$$

Therefore, the differential elastic scattering cross-section for the lth partial wave,

$$\frac{d\sigma_{sc}^l}{d\Omega} = \mid f_l(\theta,\phi)\mid^2 = \frac{(2l+1)^2}{4k^2}\mid \eta_l-1\mid^2 \{P_l(cos\theta)\}^2 \qquad (1.71)$$

And the elastic scattering cross-section, for the lth partial wave is given by

$$
\begin{aligned}
\sigma_{sc}^l &= \int_\Omega \frac{d\sigma^l}{d\Omega}d\Omega \\
&= \frac{1}{4k^2}(2l+1)^2\mid \eta_l-1\mid^2 \int_\Omega \{P_l(cos\theta)\}^2 sin\theta d\theta d\phi \\
&= \frac{1}{4k^2}(2l+1)^2\mid \eta_l-1\mid^2 \{2\pi.2/(2l+1)\} \\
&\qquad [using \ orthogonality \ of \ Legendre \ polynomial] \\
or, \ \sigma_{sc}^l &= \frac{\pi}{k^2}(2l+1)\mid 1-\eta_l\mid^2 \qquad (1.72)
\end{aligned}
$$

Since, $\mid \eta_l \mid \leq 1$, we have

$$\sigma_{sc}^l \leq \frac{\pi}{k^2}(2l+1)$$

So, when there is no reaction, we can write

$$\eta_l = e^{2i\delta_l}$$

where δ_l is a real phase factor. Hence,

$$
\begin{aligned}
\sigma_{sc}^l &= \frac{\pi}{k^2}(2l+1) \mid 1 - \eta_l \mid^2 \\
&= \frac{\pi}{k^2}(2l+1) \mid 1 - e^{2i\delta_l} \mid^2 \\
&= \frac{\pi}{k^2}(2l+1).2(1 - cos2\delta_l) \\
or,\ \sigma_{sc}^l &= \frac{4\pi}{k^2}(2l+1)sin^2\delta_l
\end{aligned}
\tag{1.73}
$$

Which for all allowed partial is

$$\sigma_{sc} = \frac{\pi}{k^2}\sum_{l=0}^{\infty}(2l+1) \mid 1 - \eta_l \mid^2 \tag{1.74}$$

$$or,\ \sigma_{sc}^l = \frac{4\pi}{k^2}\sum_{l=0}^{\infty}(2l+1)sin^2\delta_l \tag{1.75}$$

Derivation of reaction cross-section (σ_{re}) formula

The total wave function in the presence of scatterer i.e. the target nucleus can be expressed as the sum of the spherical outgoing and spherical incoming waves of different angular momentum l, so that,

$$\psi(\vec{r}) = \psi(r,\theta) = \sum_l \psi_{in}^l + \sum_l \psi_{out}^l \tag{1.76}$$

where

$$\psi_{in}^l = \frac{1}{2ikr}(2l+1)i^l P_l(cos\theta)e^{-i(kr-l\pi/2)} \tag{1.77}$$

$$\psi_{out}^l = \frac{1}{2ikr}(2l+1)i^l P_l(cos\theta)\eta_l e^{i(kr-l\pi/2)} \tag{1.78}$$

The radial part of the above waves are respectively

$$u_{in}^l = \frac{1}{2ikr}(2l+1)i^l e^{-i(kr-l\pi/2)} \tag{1.79}$$

$$u_{out}^l = \frac{1}{2ikr}(2l+1)i^l \eta_l e^{i(kr-l\pi/2)} \tag{1.80}$$

As the probability current density is defined as

$$J = \frac{i\hbar}{2m}[\psi\nabla\psi^* - \psi^*\nabla\psi] \qquad (1.81)$$

We have

$$J_{in}^l = \frac{i\hbar}{2m}[u_{in}^l\frac{\partial u_{in}^{l*}}{\partial r} - u_{in}^{l*}\frac{\partial u_{in}^l}{\partial r}] = -\frac{\hbar(2l+1)^2}{4mkr^2} \qquad (1.82)$$

$$J_{out}^l = \frac{i\hbar}{2m}[u_{out}^l\frac{\partial u_{out}^{l*}}{\partial r} - u_{out}^{l*}\frac{\partial u_{out}^l}{\partial r}] = -\frac{\hbar(2l+1)^2}{4mkr^2}\mid\eta_l\mid^2 \qquad (1.83)$$

Hence, the net outgoing current density for the lth partial wave will be

$$J^l = J_{in}^l - J_{out}^l = -\frac{\hbar(2l+1)^2}{4mkr^2}(1-\mid\eta_l\mid^2) \qquad (1.84)$$

Now, if N be the flux of the incident particles then the reaction cross-section for the lth partial wave will be

$$\sigma_{re}^l = N_{re}/N = N_a/N \qquad (1.85)$$

where $N_{re} = N_a$ is the total number of particles absorbed in going through a spherical surface of radius r:

$$N_a = -\int\int J^l r^2 d\Omega\{P_l(cos\theta)\}^2$$

the minus sign is introduced to make the net number of particles positive. We have

$$N_a = \frac{\hbar(2l+1)^2}{4mk}(1-\mid\eta_l\mid^2)\int\int\{P_l(cos\theta)\}^2\frac{r^2sin\theta d\theta d\phi}{r^2}$$

$$= \frac{\pi\hbar(2l+1)}{mk}(1-\mid\eta_l\mid^2) \qquad (1.86)$$

The incident flux is

$$N = v\mid\psi_{inc}\mid^2 = v = p/m = \hbar k/m \qquad (1.87)$$

Using (1.75) and (1.76) in (1.74) we get,

$$\sigma_{re}^l = N_a/N = \frac{\pi(2l+1)}{k^2}(1-\mid\eta_l\mid^2) \qquad (1.88)$$

Which for all allowed partial waves will be

$$\sigma_{re} = \frac{\pi}{k^2}\sum_{l=0}^{\infty}(2l+1)(1-\mid\eta_l\mid^2) \qquad (1.89)$$

Total partial cross-section and its limiting values

The total cross section i.e. the sum of the scattering and reaction cross section is

$$
\begin{aligned}
\sigma_t^l &= \sigma_{sc}^l + \sigma_{re}^l \\
&= \frac{\pi(2l+1)}{k^2}\mid 1 - \eta_l\mid^2 + \frac{\pi(2l+1)}{k^2}(1-\mid\eta_l\mid^2)
\end{aligned}
\qquad (1.90)
$$

i) For $\eta_l = 1$, both $\sigma_{sc}^l = 0$ and $\sigma_{re}^l = 0$, so $\sigma_t^l = 0$.

ii)For $\eta_l = 0$, both $\sigma_{sc}^l = \frac{\pi(2l+1)}{k^2}$ and $\sigma_{re}^l = \frac{\pi(2l+1)}{k^2}$, so $\sigma_t^l = \frac{2\pi(2l+1)}{k^2}$.

iii) For $\eta_l = -1$, $\sigma_{sc}^l = \frac{4\pi(2l+1)}{k^2}$ and $\sigma_{re}^l = 0$, so $\sigma_t^l = \sigma_{sc}^l(max) = \frac{4\pi(2l+1)}{k^2}$

Thus, we may conclude that there can be elastic scattering without reaction, but reaction cannot occur without scattering. For reaction to be possible $-1 < \eta_l < 1$.

1.3   Particle physics

Particle physics (also called high energy physics) is the branch of physics that deals with the nature of the particles that constitute matter (particles with mass) and radiation (mass less particles). Although the word "particle" can refer to various types of very small objects (e.g. protons, gas particles, or even household dust), "particle physics" usually investigates the irreducibly smallest detectable particles and the irreducibly fundamental force fields necessary to explain them. By our current understanding, these elementary particles are excitations of the quantum fields that also govern their interactions. The currently dominant theory explaining these fundamental particles and fields, along with their dynamics, is called the Standard Model. Thus, modern particle physics generally investigates the Standard Model and its various possible extensions, e.g. to the newest "known" particle, the Higgs boson, or even to the oldest known force field, gravity. In nature, there are four fundamental forces or interactions through which particles can interact: the strong, weak, electromagnetic and gravitational interactions. There are exchange particles associated with each of the fundamental forces: the photon (for electromagnetism), the W^+, W^- and Z_0 particles (for the weak interaction), and eight kinds of gluon (for the strong interaction), all being spin 1 particles. The natural forces, their relative strength and the particles affected by each are summarized in Table 1.3. The strong, electromagnetic and weak interactions of the fundamental particles (and their antiparticles) may be described theoretically by quantum field theories in which the forces

Table 1.3: Forces of nature, their relative strengths, associated particles and quanta.

Interaction	Relative strength	Range	Particles affected	Exchanged quanta
Electromagnetic	1	Long range	Charged particles	Photons
Strong	10	Short range $(10^{-15}m)$	Quarks	Gluons
Weak nuclear	10^{-12}	Very short range $(10^{-18}m)$	Quarks, leptons	$W^{\pm}, Z_0$ bosons
Gravitational	10^{-36}	Long range	Particles with mass	Graviton (undetected)

are mediated by exchange particles. These quantum field theories have been developed into the standard model of particle physics. This is not thought to be the final theory, but its construction is regarded as a major triumph since, by using Feynman diagrams and other techniques, it permits the evaluation of measurable quantities such as cross-sections and mean lifetimes. Going beyond the standard model might require the formulation of a grand unified theory, or a string theory that might involve super symmetry. Today, it is believed that, there are two families of fundamental particles, called leptons and quarks. By fundamental we mean that there is no evidence that these particles are composed of smaller or simpler constituents. There are just six leptons and six quarks, together with an equal number of their antiparticles. All the familiar forms of matter are ultimately composed of these particles. The six different types are often referred to as different flavors of lepton, and the three pairs are said to represent the three generations of leptons. The first generation consists of the familiar electron (e) and its β-decay partner, the electron neutrino (ν_e). The second generation of leptons consists of the muon (μ) and a second type of neutrino called a muon neutrino (ν_μ). The muon is similar to the electron except that it is about 200 times heavier and unstable with a fairly long lifetime of a few microseconds. The third generation of leptons consists of a particle called a tauon (τ) and a third type of neutrino called a tauon neutrino (ν_τ). The tauon is similar to and even heavier than the muon and has a much shorter lifetime. These two heavier leptons, being unstable, are not normally constituents of matter, but are created in high-energy collisions between other subatomic particles. Associated with these six leptons are the six anti-leptons, particles of antimatter. These include the positron (e^+) which is the antiparticle of the electron and the electron antineutrino $(\overline{\nu_e})$. Leptons are spin 1/2 particles and have lepton number L = 1 whilst the corresponding family of 6 anti-leptons have L = -1. Leptons feel the weak interaction and the charged ones also feel the electromagnetic interaction, but leptons do not feel the strong interaction. The pattern of the leptons is repeated for the quarks. The six types (or flavors) of quark are labeled (for historical reasons) by the letters u, d, c, s, t and b, which stand for up, down, charm, strange, top and bottom. Like the leptons, the quarks are paired off in three generations on the basis of their mass. To each quark, there corresponds an anti-quark, with the opposite electric charge and the same mass. The antiquarks are denoted by $\overline{u}, \overline{u}, \overline{d}, \overline{c}, \overline{s}, \overline{t}$ *and* $\overline{b}$. Unlike leptons, the quarks and antiquarks have never been observed in isolation. They only seem to occur in bound together in combinations held together by gluons. For example, the familiar proton is a combination of two up quarks and a down quark, which we can write as uud. Note that each up quark carries an electric charge of $+2e/3$ and a down quark carries a charge of $-e/3$, so the combination uud does indeed give a net electric charge equal to the charge e on a proton. Similarly, a neutron is the combination udd which has a net electric charge of zero. Observable particles consisting of combinations of quarks are collectively called hadrons and there are literally hundreds of them, the proton and neutron being the most familiar. There are three recipes for building hadrons from quarks: A hadron can consist of: three quarks (in which case it is called a baryon); three antiquarks (in which case it is called an antibaryon); or one quark and one anti quark (in which case it is called a meson). Baryons have half odd-integer spin (1/2, 3/2, etc.) and baryon number B = 1 whilst the corresponding family of anti baryons have B = -1. Mesons have zero or integer spin (0, 1, 2, etc.) and baryon number B = 0. Baryons and mesons can interact through the strong interaction as well as the weak interaction and, if charged, the electromagnetic interaction. When particles collide, certain quantities such as total electric charge, total (relativistic) momentum and total (relativistic) energy are always conserved. The particles may undergo elastic collisions where the colliding particles simply exchange energy and momentum, or inelastic collisions where (relativistic) kinetic energy may be converted into rest energy and so the nature and number of particles may change. The likelihood of a particle being scattered in a given process at a specified energy is described by a measurable quantity called a cross-section. Cross-sections are measured in barns (1 barn = $10^{-28}m^2$).

1.4   Astrophysics

Astrophysics is a branch of space science that applies the laws of physics and chemistry to explain the birth, life and death of stars, planets, galaxies, nebulae and other objects in the universe. It has two main science components- astronomy and cosmology. It can also be categorized as application of physics to the art of observation of astronomical objects by using radio, optical, IR, UV, x-ray, γ-ray instruments. And also neutrino, gravity-wave studies, measurements of positions, bright nesses, spectra, structure of gas clouds, planets, stars, galaxies, globular clusters, clusters of galaxies, super-clusters, quasars, etc. for better understanding and interpretation of the facts. Some of the basic characteristics of astrophysics are:

- Large range of scales: from nuclear scales ($1fm = 10^{-15}m = 10^{-13}cm$) to cosmological scales ($10Gpc = 10^{10}pc = 3 \times 10^{26}m = 3 \times 10^{28}cm$) having a range $\sim 3 \times 10^{41}$. Some physical problems involve large scale while some involve small scales; for example, the Chandrasekhar mass, $M_{Ch} \simeq m_p^{-2}(\hbar c/G)^{3/2} = 3.7 \times 10^{33}g = 2M_\odot$ involves small-scale physics ($\hbar$) and yet describes something with a mass about twice the mass of the Sun.

- Complexity of systems: Systems are often so complicated that we cannot always conduct experiments to isolate the relevant variables /parameters involved in the observables. Data accumulated by astronomical observations in most cases are far from accuracy and incomplete so one often restricts himself in order of magnitude estimation only.

- Versatility of subjects: Astrophysics covers wide range of physics which sometimes may include topics beyond the normal physics curriculum: e.g., fluid mechanics, magneto-hydrodynamics, radiative transfer, etc. Phenomena like flux freezing, gravitational collapse, etc., that cannot be realized in the laboratory may also occur in astrophysics, so astrophysical interpretations may sometimes offer only a little part of physics.

- Restriction due to universality of physical laws: Astrophysics does not allow new physics to be involved very often as it corresponds to an application of the standard laws of physics to the Universe as a whole. It assumes the universality (literally) of the laws of physics in order to make any progress at all and to be allowed to call the subject astrophysics. The evidence for this point of view may be derived from the laboratory laws to the extent the reach the distant objects in the Universe and provide explanations of phenomena there.

- Connection to astronomy: Astrophysics is intimately connected with what astronomy is able to observe. Until recently, there was little high-energy astrophysics because there was no high-energy (x-ray and γ-ray) astronomy. Now there is, and we can see much hotter parts of objects than we knew about before.

- Assumption of Copernican principle: Astronomy and astrophysics generally assumes the Copernican principle based on applicable physical laws and not just geometrical arguments and logic.

1.5   Scopes of nuclear astrophysics

Nuclear physics has found an incredible number of applications and connections with other fields in one century, although, in the narrowest sense, it is only concerned with bound systems of protons and neutrons. From the beginning however, progress in the study of such systems was possible only because of progress in the understanding of other particles: electrons, positrons, neutrinos and, eventually quarks and gluons. In fact, we now have a more complete theory for the physics of these "elementary particles" than for nuclei as such. A nuclear species is characterized by its number of protons Z and number of neutrons N . There are thousands of combinations of N and Z that lead to nuclei that are sufficiently long-lived to be studied in the laboratory. The large number of possible combinations of neutrons and protons is to be compared with the only 100 or so elements characterized simply by Z. A "map" of the galaxy of nuclei is shown in Figure 1.9. Most nuclei are unstable, i.e. radioactive. Generally, for each A = N + Z there is only one or two combinations of (N, Z) sufficiently long-lived to be naturally present on Earth in significant quantities. These nuclei are indicated by the black squares in Figure 1.9 and define the bottom of the valley of stability in the figure. One important line of nuclear research is to create new nuclei, both high up on the sides of the valley and, especially, super-heavy nuclei beyond the heaviest now known with A = 292 and Z = 116. Phenomenological arguments suggest that there exists an "island of stability" near Z = 114 and 126 with nuclei that may be sufficiently long-lived to have practical applications.

The physics of nuclei as such has been a very active domain of research in the last thirty years owing to the construction of new machines, the heavy

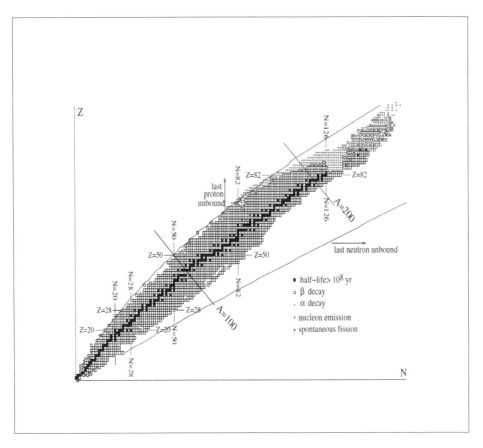

Figure 1.9: The nuclei. The black squares are long-lived nuclei present on Earth. Unbound combinations of (N, Z) lie outside the lines marked "last proton/neutron unbound." Most other nuclei β-decay or α-decay to long-lived nuclei

ion accelerators of Berkeley, Caen (GANIL), Darmstadt and Dubna. The physics of atomic nuclei is in itself a domain of fundamental research. It constitutes a true many-body problem, where the number of constituents is too large for exact computer calculations, but too small for applying the methods of statistical physics. In heavy ion collisions, one discovers subtle effects such as local super fluidity in the head-on collision of two heavy ions. Nuclear physics has had an important by-product in elementary particle physics and the discovery of the elementary constituents of matter, quarks and leptons, and their interactions. Nuclear physics is essential to the understanding of the structure and the origin of the world in which we live. The birth of nuclear astrophysics is a decisive step forward in astronomy and in cosmology. Perhaps the most important achievement of nuclear physics is its ability to explain the mechanism behind the production of enormous energy and nucleosynthesis in the stellar environment. Astrophysicists now believe that they understand in some detail life cycle of the stellar objects including their birth as diffuse clouds and death as white dwarfs or supernovae. Through this process, nuclear physics has allowed us to understand how the initial mix of hydrogen and helium produced in the primordial Universe has been transformed into the interesting mixture of heavy elements that makes terrestrial life possible. Studies of the atomic nucleus offer profound insights into the nature of matter, their origin, and energy content. According to modern science, the matter we see around us is composed of different types of atoms characterizing the elements. Deep in the heart of each atom there is the nucleus, which is composed of yet smaller particles, protons and neutrons together called nucleons. How they behave is controlled by three fundamental forces of nature -the strong force, together with the weak and electromagnetic forces. **These natural forces combine to generate highly complex nuclear structures that are challenging subject of study for clear understanding, primarily based on some important scientific questions on i) the arrangements of nucleons in nuclei, ii) types of nuclei and factors affecting their stability, iii) binding mechanism of quarks in the nucleons, iv) formation of nucleons from quarks, v) formation of elements from nucleons, vi) properties and behaviors of nuclei etc.**

As already stated nuclear astrophysics is the study of the nuclear reactions that energize the sun and other stars across the cosmic universe, and also create the variety of atomic nuclei. **Almost all the elements found on Earth – except the very lightest ones - were exclusively made in stars. Understanding the underlying astrophysical processes gives us clues about: i) the origin of the elements and their rela-**

tive abundances, ii) the origin of the Earth and its composition, iii) the evolution of life, iv) the evolution of stars, galaxies and the Universe itself, v) the fundamental laws and building blocks of Nature. During their lives, stars generate many exotic, short-lived nuclei that do not exist on Earth, but are nevertheless significant in understanding the structure of all nuclear species and the fundamental forces governing them. In turn, this knowledge is essential to developing, for example, new types of safe energy and isotopes of industrial or medical use. Nuclear astrophysicists pursue their research in several ways:

- by detecting and analyzing emissions from stars, the dusty remnants from exploded stars and from compact dead' stars.

- by designing laboratory experiments that explore stellar nuclear reactions in the Big Bang, in stars and in supernova explosions.

- by analyzing geological samples and those from extraterrestrial sources such as meteorites and the grains of 'stardust[4]

- by carrying out theoretical calculations on nuclear behavior and its interplay with the stellar environment.

There have been many research successes in recent years but there are still many mysteries to solve. Astrophysics community round the globe is trying to unveil the mysteries behind this amazingly mysterious universe.

Solved Problems

1. An electron is confined to the interior of a hollow spherical cavity of radius R with impenetrable walls. Find an expression for the pressure exerted on the walls of the cavity by the electron in its ground state.

   Solution: If we suppose that the radius of the cavity increases dR. Then the work done by the electron in the process is $PdV = 4\pi R^2 P dR$, which causes a decrease of its energy by dE. Hence the pressure exerted by the electron on the walls is

$$P = -\frac{1}{4\pi r^2}\frac{dE}{dR}$$

---

[4]There exists cosmic dust in outer space having grain size between a few molecules to 0.1 μm. A smaller fraction of all dust in space consisting refractory minerals condensed as matter left by the stars is called "stardust".

For the electron in its ground state, the angular momentum $l = 0$ and the wave function has the form

$$\phi(r) = \frac{1}{\sqrt{4\pi}} \zeta(r)/r$$

where $\zeta(r)$ is the solution of the radial Schrödinger equation,

$$\frac{d^2\zeta}{dr^2} + k^2\zeta(r) = 0,$$

with $k^2 = 2mE/\hbar^2$ and $\zeta(r) = 0$ at r = 0. Thus

$$\zeta(r) = A \sin kr$$

As the walls cannot be penetrated, $\zeta(r) = 0$ at r = R, giving $k = \pi/R$. Hence the energy of the electron in the ground state is

$$E = \frac{\pi^2\hbar^2}{2mR^2},$$

and the pressure is

$$P = -\frac{1}{4\pi r^2}\frac{dE}{dR} = \frac{\pi\hbar^2}{4mR^5}$$

2. The muon is a relatively long-lived elementary particle with mass 207 times the mass of electron. The electric charge and all known interactions of the muon are identical to those of the electron. A "muonic atom" consists of a neutral atom in which one electron is replaced by a muon. (a) What is the binding energy of the ground state of muonic hydrogen? (b) What ordinary chemical element does muonic lithium (Z = 3) resemble most? Explain your answer.

Solution: (a) By analogy with the hydrogen atom, the binding energy of the ground state of the muonic atom is

$$E_0 = \frac{m_\mu e^4}{2\hbar^2} = 207E_H = 2.82 \times 10^3 eV.$$

(b) A muonic lithium atom behaves chemically most like a He atom. As μ and electron are different fermions, they fill their own orbits. The two electrons stay in the ground state, just like those in the He atom, while the μ stays in its own ground state, whose orbital radius is 1/207 of that of the electrons. The chemical properties of an atom is determined by the number of its outer most shell electrons. Hence the mesic atom behaves like He, rather than like Li.

3. What is the density of nuclear matter in ton/cm^3?

   Solution: The linear size(r_n) of a nucleon is about $10^{-13}cm$, so the volume (v_n) of a nucleon is about $10^{-39}cm^3$. The mass of a nucleon (m_n) is about $10^{-27}kg = 10^{-30}ton$, so the density (ρ_n) of nuclear matter is

$$\rho_n = m_n/V_n \simeq \frac{10^{-30}}{10^{-39}} = 10^9 ton/cm^3.$$

4. (a) Calculate the electrostatic energy of a charge Q distributed uniformly throughout a sphere of radius R. (b) Since $^{27}_{14}$Si and $^{27}_{13}$Al are "mirror nuclei", their ground states are identical except for charge. If their mass difference is 6 MeV, estimate their radius (neglecting the proton-neutron mass difference).

   Solution: (a) The electric field intensity at the centre of uniformly charged sphere is

$$\begin{aligned} E(r) &= \frac{Qr}{R^3}\ for\ r < R \\ &= \frac{Q}{r^2}\ for\ r > R. \end{aligned}$$

The electrostatic energy is

$$\begin{aligned} W &= \int_0^\infty \frac{E^2}{8\pi}dV \\ &= \int_0^R \frac{(Qr/R^3)^2}{8\pi}4\pi r^2 dr + \int_R^\infty \frac{(Q/r^2)^2}{8\pi}4\pi r^2 dr \\ &= \frac{Q^2}{8\pi}\left[\int_0^R (\frac{r}{R^3})^2 4\pi r^2 dr + \int_R^\infty (\frac{1}{r^2})^2 4\pi r^2 dr\right] \\ &= \frac{Q^2}{2}\left[\int_0^R \frac{r^4}{R^6}dr + \int_R^\infty \frac{1}{r^2})dr\right] \\ &= \frac{Q^2}{2}(\frac{1}{5R} + \frac{1}{R}) \\ &= \frac{3Q^2}{5R} \end{aligned}$$

(b) The mass difference between the mirror nuclei $^{27}_{14}$Si and $^{27}_{13}$Al can be considered as due to the difference in their electrostatic energy:

$$\Delta W = \frac{3e^2}{5R}(Z_1^2 - Z_2^2).$$

$$
\begin{aligned}
R &= \frac{3e^2}{5\Delta W}(Z_1^2 - Z_2^2) \\
&= \frac{3e^2}{5\Delta W}(14^2 - 13^2) \\
&= \frac{3c\hbar}{5\Delta W}\left(\frac{e^2}{c\hbar}\right)(14^2 - 13^2) \\
&= \frac{3 \times 1.97 \times 10^{-11}}{5 \times 6}\left(\frac{1}{137}\right)(14^2 - 13^2)cm \\
&= 3.88 \times 10^{-11}cm \\
&= 3.88 fm
\end{aligned}
$$

5. The semi empirical mass formula modified for nuclear-shape eccentricity suggests a binding energy for the nucleus $^{A}_{Z}\mathrm{X}$

$$
B = \alpha A - \beta A^{2/3}\left(1 + \frac{2}{5}\epsilon^2\right) - \gamma Z^2 A^{-1/3}\left(1 - \frac{1}{5}\epsilon^2\right)
$$

where $\alpha, \beta, \gamma = 14, 13, 0.6 MeV$ and ϵ is the eccentricity. Briefly interpret this equation and find a limiting condition involving Z and A such that a nucleus can undergo prompt (unhindered) spontaneous fission. Consider $^{238}_{92}\mathrm{U}$ and $^{240}_{94}\mathrm{Pu}$ as examples.

Solution: In the mass formula, the first term represents volume energy, the second term surface energy, in which the correction $\frac{2}{5}\epsilon^2$ is for deformation from spherical shape of the nucleus, the third term, the Coulomb energy, in which the correction $\frac{1}{5}\epsilon^2$ is also for nucleus deformation. Consequent to nu clear shape deformation, the binding energy is a function of the eccentricity ϵ. The limiting condition for stability is

$$
\frac{dB}{d\epsilon} = 0.
$$

We have

$$
\frac{dB}{d\epsilon} = -\frac{4}{5}\beta A^{2/3}\epsilon + \frac{\gamma Z^2}{A^{1/3}}\left(\frac{2}{5}\epsilon\right) = \frac{2}{5}\epsilon A^{2/3}\left(\frac{\gamma Z^2}{A} - 2\beta\right).
$$

If $\frac{dB}{d\epsilon} > 0$, nuclear binding energy increases with ϵ so the deformation will keep on increasing and the nucleus becomes unstable. If $\frac{dB}{d\epsilon} < 0$, binding energy decreases as ϵ increases so the nuclear shape will tend to that with a lower ϵ and the nucleus is stable. So the limiting condition for the nucleus to undergo prompt spontaneous fission is $\frac{dB}{d\epsilon} > 0$, or

$$
\frac{Z^2}{A} \geq \frac{2\beta}{\gamma}.
$$

For given set of data the fraction

$$\frac{2\beta}{\gamma} = 2.0 * 13/0.6 = 43.33$$

For $^{238}_{92}$U, $Z^2/A = 35.56 < 43.3$, so it cannot undergo prompt spontaneous fission; it has a finite lifetime against spontaneous fission. For $^{240}_{94}$Pu, $Z^2/A = 36.8 < 43.3$, it is also stable against prompt spontaneous fission.

6. Give examples of the three nuclear reactions highly considered for controlled thermonuclear fusion. Which among them has the largest cross section? Give the approximate energies released in the reactions. How would any resulting neutrons be used?

   Solution: Reactions often considered for controlled thermonuclear fusion are

$$
\begin{aligned}
D + D &\rightarrow {}^{H}e + n + 3.25 MeV, \\
D + D &\rightarrow T + p + 4.0 MeV, \\
D + T &\rightarrow {}^{4}He + n + 17.6 MeV.
\end{aligned}
$$

The cross section of the last reaction is the largest. Neutrons resulting from the reactions can be used to induce fission in a fission-fusion reactor, or to take part in reactions like $^{6}Li + n \rightarrow^{4} He + T$ to release more energy.

Exercise

1. If the nuclear force is charge independent and a neutron and a proton form a bound state, then why is there no bound state for two neutrons or two protons? What information do they provide on the nucleon-nucleon force?

2. What is the typical cross section for low energy neutron-nucleus scattering ? [Hints:$\sim 4\pi R^2$,R is the radius of sphere of action of nuclear force.]

3. Having 4.5 GeV free energy, what is the most massive isotope one could theoretically produce from nothing? Can you identify such isotope from among ^{2}D, ^{3}He and ^{3}T? [Ans. ^{2}D]

4. The beta decay half-life of a radioactive specimen is 1s and that for alpha decay is 3s. What is the half-life of the radioactive specimen? [0.75s]

References

Quoted in text:

- E. M. Burbidge; G. R. Burbidge; W. A. Fowler and F. Hoyle. (1957). "Synthesis of the Elements in Stars". Reviews of Modern Physics 29 (4): 547.

- Cameron, A.G.W. (1957). Stellar Evolution, Nuclear Astrophysics, and Nucleogenesis (Report). Atomic Energy of Canada.

- Barnes, C. A.; Clayton, D. D.; Schramm, D. N., eds. (1982), Essays in Nuclear Astrophysics, Cambridge University Press, ISBN 0-52128-876-2

- Henri Becquerel (1896). "Sur les radiations émises par phosphorescence". Comptes Rendus 122: 420–421.

- Massimo S. Stiavelli. From First Light to Reionization. John Wiley & Sons, Apr 22, 2009. Pg 8.

- M. Junker et al., Phys. Rev. C 57 (1998) 2700.

- A. Krauss et al., Nucl. Phys. A467 (1987) 273.

- D. Zahnow et al., Z. Phys. A 351 (1995) 229-236

Further readings

- *Nuclear Structure* Bohr, A. & Mottelson, B.:Benjamin, New York, 1969

- *Structure of the Nucleus* Preston, M. A. & Bhaduri, R. K.:Addison-Wesley, Reading, 1975.

- *Nuclear Physics* Wong, S. M.:John Wiley, New York, 1988.

- *Particle Physics- A Comprehensive Introduction* Abraham Seiden: Pearson

- *Theoretical Astrophysics* Padmanabhan, T.: Cambridge University Press

Chapter 2

Nucleosynthesis

2.1 Introduction

Apart of nuclear fusion, another important keyword of nuclear astrophysics is nucleosynthesis. It is responsible for the production of the chemical elements that constitute the baryonic matter of the universe by the process of thermonuclear fusion reactions. The nature of distribution of chemical elements in the sun and other stars that belongs to the main sequence have relatively larger abundances of lighter elements like hydrogen, helium etc compared to the heavier elements. Plot of relative abundances of elements against their mass number (A) (Figure 2.1) shows rapid decrease in its value with increasing mass number (A) for smaller A corresponding to the lighter elements and becomes almost constant for A exceeding 100 along with some local ups and downs. For instance, the abundance of lighter elements like deuterium(^{2}H), lithium (^{6}Li), beryllium (^{8}Be) and boron (^{10}B) is unusually low, while that for iron (^{56}Fe) is unusually large indicated by a narrow peak at A=56. The plot also exhibits some double peaks corresponding to A=80, 90; 130, 138; 196, 208 in the neighborhood of neutron magic numbers 50, 82, 126 respectively. Thus, theory of nucleosynthesis must account for the observed features of abundance curve. It can be said that nuclear processes, operating both in the early universe and in stars, are responsible for the synthesis of the elements. The history of nuclear matter is composed of diverse constituents like stars, interstellar (and intergalactic) gas and dust, meteorites, and cosmic rays. Rapid advancements in observational techniques and experimental facilities (e.g. the Hubble Space Telescope, the Compton Gamma-Ray Observatory, the Rosat X-ray Observatory, the Keck telescopes) have made it possible to determine accurately elemental and iso-

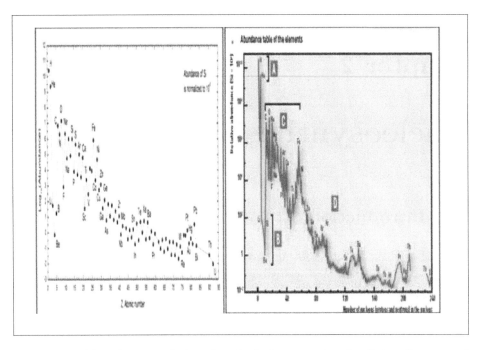

Figure 2.1: Abundances of the chemical elements in the Solar System. A: The most abundant elements are hydrogen (^{1}H) and helium (^{4}He) constituting on average 98% of the matter of the entire observable universe. B: The next three elements lithium (Li), beryllium (Be), boron (B) are rare because they are poorly synthesized in the Big Bang and also in stars thereby giving rise to a significant gap between helium and carbon. C: The most abundant nuclei after that are carbon with 12 nucleons (^{12}C), oxygen (^{16}O), neon (^{20}Ne), magnesium (^{24}Mg), silicon (^{28}Si) and iron $^{(56)}$Fe which are also the most stable nuclei in the universe. D: Nuclei heavier than iron (^{56}Fe). Iron is the most stable nucleus in the universe. Beyond boron (B), two general trends in the remaining stellar-produced elements are observed: (1) there is an alternation of abundance of elements according to whether they have even or odd atomic numbers, and (2) a general decrease in abundance, as elements become heavier. Within this trend is a peak at abundances of iron and nickel, which is especially visible on a logarithmic graph (left figure).

topic abundances in many astronomical environments. It is to be noted that unlike other cases, the cases of earth and other minor planets nearer the sun, the relative abundance of lighter elements like hydrogen and helium are very small. This is because of the fact that at higher temperatures the relatively weak gravitational fields were unable to prevent the escape of these elements (due to their large RMS speed from the view point of kinetic theory of gases) from their atmosphere. But in larger planets like Jupiter, Saturn etc located far away from the sun has sufficient gravitational field strengths to prevent the escape of these lighter elements from their atmosphere at lower temperatures. So these elements are still found in the atmosphere of these planets in relatively greater abundance. Further if we analyze the chemical composition of the Earth's crust, it will be found that, it is mainly composed of stable even-even nuclei which accounts for more than 86%. Among these, doubly magic nuclei constitute more than 50%. For examples we may take oxygen ($^{16}_{8}O$ isotope 99.7%), silicon ($^{28}_{14}Si$ isotope 92.3%), calcium ($^{40}_{20}Ca$ isotope 97%). Thus we have clear evidences that theory of nucleosynthesis is correlated with the properties of the nuclei, since the even-even nuclei constitute more than 60% of the total stable isotopes and doubly magic nuclei being more stable have still larger abundances. Such studies impose increasingly stringent constraints on models of stellar evolution and nucleosynthesis, which in turn help identifying critical areas where input of nuclear physics is unavoidable. Rigorous numerical models of nucleosynthesis have played crucial roles to identify promising sites for the various processes. In the pre-galactic nucleosynthesis age, the big bang created the light nuclei ^1H, ^2H, ^3He, ^4He, and ^7Li. Reactions between nuclei in the interstellar gas and high energy cosmic rays also form these elements in the way interactions of neutrinos with heavier nuclei in supernovae. Some of the heavier elements at concentration much less (10^{-4} times) than that of the sun may have been formed prior to the formation of galaxies. The synthesis of the bulk of the heavy nuclei present in galactic matter, however, is generally attributed to processes in stellar and supernova environments, following the formation of galaxies. Helium burning stars with $M_s < M < 10M_\odot$ are the predominant source of carbon, nitrogen, and roughly half the nuclei heavier than iron. Massive stars ($M > 10M_\odot$) and the associated supernovae (SNe II) produce most of the nuclear species from oxygen through zinc and a significant fraction of the heavier nuclei. Nuclei up to calcium are made primarily in the stable burning pre-supernova phases, while the iron-peak nuclei are produced in explosive silicon burnings. Type Ia supernovae produce a substantial fraction of the iron peak nuclei and some intermediate mass isotopes in explosive carbon, oxygen, and silicon burning processes. In addition to the above, abundance constraints, some observational constraints are also imposed on nucleosynthesis: Big bang synthesis of the light nuclei ^{1}H, ^{2}H, ^{3}He, ^{4}He, and ^{7}Li. is constrained by the abundances of these isotopes observed in our galaxy and other galaxies, and in distant gas clouds. Cosmic ray spallation[1] synthesis of the light isotopes ^{6}Li, ^{7}Li, ^{9}Be, ^{10}B, and ^{11}B is constrained by recent studies of lithium, beryllium, and boron abundances in old metal poor stars in the galactic halo.

Nucleosynthesis via the CNO, NeNa, and MgAl burning cycles is reflected in abundances in globular cluster giants. For example, significant depletions of oxygen are anti correlated with the abundances of nitrogen, sodium, and aluminum. These observations bear on the nucleosynthetic and convective processes in massive stars.

Successive phases of carbon, neon, and oxygen burning, associated with massive star and Type II supernova environments, are responsible for the enhancements of the α-elements O, Ne, Mg, etc. compared to the Fe observed in old stars located in the halo of our galaxy. This overproduction of α-elements in Type II supernovae must ultimately be compensated for by other processes, presumably with material rich in Fe from Type Ia supernovae. Nucleosynthesis processes in novae are constrained by the abundances of many elements with masses A $<$ 35, found in nova ejects.

Recent studies of extremely metal-deficient halo stars find r-process abundances in the mass range from barium through the actinides that are closely proportional to solar system r-process abundances. Data on some of the same stars show that r-process nuclei with A$<$130 are not produced in solar proportions relative to the heavy r-process elements. This could mean that r-process production of light nuclei is sensitive to the details of a specific site, or it may be evidence that multiple r-process sites are contributing. Nuclei in the iron-peak region are probably the products of explosive nucleosynthesis in core-collapse and Type Ia supernovae. Observations of gamma rays from Supernova 1987A revealed that approximately $0.075 M_\odot$ of ^{56}Ni was ejected with an isotopic ratio $^{57}Ni/^{56}Ni$ that is approximately 1.5 times that seen in the solar system. Gamma rays from ^{44}Ti have been detected from the remnant of the Cas A supernova. These results can be used to determine how much material was ejected into the interstellar medium (compared to that accreted on the supernova remnant) by a supernova explosion, as well as the

---

[1]Spallation is a process in which fragments of material (spall) are ejected from a body due to impact or stress. In planetary physics, spallation describes meteoritic impacts on a planetary surface and the effects of a stellar wind on a planetary atmosphere. In nuclear physics, spallation is the process in which a heavy nucleus emits a large number of nucleons as a result of being hit by a high-energy particle, thus greatly reducing its atomic weight.

conditions deep inside the exploding star. A variety of sources can produce the ^{26}Al in the galaxy. A map of the distribution of gamma rays from ^{26}Al is shown in Figure 1.9. When more information is available for the distribution of ^{60}Fe gamma rays, a comparison of these distributions should strongly constrain the nature of the sources of ^{26}Al. Presolar grains, tiny (1.6 nm to 20 μm) inclusions found in meteorites, yield isotopic abundances corresponding to individual astrophysical events, some apparently from s-process sites and planetary nebulae, and others from supernovae. These abundances serve to test models of s-process nucleosynthesis in shell-helium burning (AGB) stars and of explosive burning in supernova shock fronts. These great advances in observation draw attention to the fact that we have far to go to reach a real understanding of the processes involved. For example: we do not know the sources of the cosmic rays, we do not understand the origin of Type Ia supernovae, we cannot consistently calculate a SNe II explosion, we do not know where the r-process takes place, and we do not yet have a convincing explanation of why we observe fewer neutrinos than expected from our sun.

The basic codes of belief of nuclear astrophysics are that only isotopes of hydrogen and helium (and traces of few more light elements like lithium, beryllium, and boron) can be formed in a homogeneous big bang model (see big bang nucleosynthesis), and all other elements are formed in stars. The conversion of nuclear mass to radiative energy (by merit of Einstein's famous mass-energy relation in relativity) is the source of energy which allows stars to shine for up to billions of years. Many notable physicists of the 19th century, such as Mayer, Waterson, von Helmholtz, and Lord Kelvin, postulated that the Sun radiates thermal energy based on converting gravitational potential energy into heat. The lifetime of the Sun under such a model can be calculated relatively easily using the virial theorem, yielding around 19 million years, an age that was not consistent with the interpretation of geological records or the then recently proposed theory of biological evolution. Some calculation indicates that if the Sun consisted entirely of a fossil fuel like coal, a source of energy familiar to many people, considering the rate of thermal energy emission, then the Sun would have a lifetime of merely four or five thousand years, which is not even consistent with records of human civilization. The now discredited hypothesis that gravitational contraction is the Sun's primary source of energy was, however, reasonable before the advent of modern physics; radioactivity itself was not discovered by Becquerel until 1895. Besides the prerequisite knowledge of the atomic nucleus, a proper understanding of stellar energy is not possible without the theories of relativity and quantum mechanics. After Aston demonstrated that the mass of helium is less than four times the mass of the proton, Eddington proposed that in the core of the Sun, through an unknown process, hydrogen was transmuted into helium, liberating energy. 20 years later, Bethe and von Weizsaecker independently derived the CN cycle, the first known nuclear reaction cycle which can accomplish this transmutation; however, it is now understood that the Sun's primary energy source are the pp-chains, which can occur at much lower energies and are much slower than catalytic hydrogen fusion. The time-lapse between Eddington's proposal and the derivation of the CN cycle can mainly be attributed to an incomplete understanding of nuclear structure, and a proper understanding of nucleosynthetic processes was not possible until Chadwick discovered the neutron in 1932 and a contemporary theory of beta decay developed.

The theory of stellar nucleosynthesis reproduces the chemical abundances observed in the solar system and galaxy, which from hydrogen to uranium, show an extremely varied distribution spanning twelve orders of magnitude (one trillion). While impressive, these data were used to formulate the theory, and a scientific theory must be predictive in order to have any merit. The theory of stellar nucleosynthesis has been well-tested by observation and experiment since the theory was first formulated. The theory predicted the observation of technetium (the lightest chemical element with no stable isotopes) in stars, observation of galactic gamma emitters such as ^{26}Al and ^{44}Ti observation of solar neutrinos, and observation of neutrinos from supernova 1987. These observations have far reaching implications. ^{26}Al has a lifetime a bit less than one million years, which is very short on a galactic timescale, proving that nucleosynthesis is an ongoing process even in our own time. Work which lead to the discovery of neutrino oscillation, implying a non-zero mass for the neutrino and thus not predicted by the Standard Model of particle physics, was motivated by a solar neutrino flux of about three times lower than expected. This was a longstanding concern in the nuclear astrophysics community such that it was colloquially known simply as the Solar neutrino problem. The observable neutrino flux from nuclear reactors is much larger than that of the Sun, and thus Davis and others were primarily motivated to look for solar neutrinos for astronomical reasons. Although the foundations of the science are genuine, there remains, still many open questions. A few of the long-standing issues are helium fusion specifically through the $^{12}C(\alpha, \gamma)^{16}O$ reaction, the astrophysical site of the r-process, anomalous lithium abundances in Population III stars, and the explosion mechanism in core-collapse supernovae. As discussed above, the evolution of the universe can divided into four stages:

I) Primordial Nucleosynthesis and Formation of Atoms, II) Galactic Condensation, III) Stellar Nucleosynthesis, and IV) Evolution

of the Solar System.

The age of the universe is the grand total duration of the above four stages counted sequentially from its inception in the Big Bang[2], a hot, dense expanding fireball. The first stage lasts from the so called Big Bang (t=0) up to the period of about 10^6 years which is a small interval in astronomical scales and hence the uncertainties involved in the estimation of this period is almost insignificant. Galactic condensation, the second stage is caused solely by the influence gravitational forces without any intervention of nuclear or particle physics. As estimated, this stage lasts from 1 - 2Gy (1Gy=10^9yr). The third stage or the era of stellar nucleosynthesis, although contributes the largest uncertainties ($\pm 2Gy$), yet results from recent nuclear reaction studies have provided results for this era in excellent agreement with independent astronomical deductions. Finally, the duration of the evolution of solar system is well known with very small uncertainties. Literature survey reveals the fact that there are 26 nuclides which are considered significant in the big bang nucleosynthesis (BBN) analysis. These are: **n, p, ^{2}H, ^{3}H, ^{3}He, ^{4}He, ^{6}Li, ^{7}Li, ^{7}Be, ^{8}Li, ^{8}B, ^{9}Be, ^{10}B, ^{11}B, ^{11}C, ^{12}B, ^{12}C, ^{12}N, ^{13}C, ^{13}N, ^{14}C, ^{14}N, ^{14}O, ^{15}N, ^{15}O, and ^{16}O.**

2.2 Primordial nucleosynthesis

The present configuration of the Universe is the result of the evolution of the primordial matter which was mainly composed, a few minutes after the Big Bang, of the two lightest elements, hydrogen and helium. It is well-known at present that all other chemical elements, which make up, for instance, all living beings and celestial objects have been produced by nuclear reactions in the heart of stars. These nuclear reactions produce the energy that powers the stars and this balances the gravitational force to prevent their collapse. In turn, the production of energy and elements explain the structure and evolution of the Universe, with stars that will progressively cool down and die and stars that will explode producing cataclysmic events. To understand the structure and the evolution of the Universe it is essential to understand the synthesis of the elements. The synthesis of heavy nuclei begins with the first reaction between proton and neutron resulting in the production of deuteron and energetic gamma photon

$$n + p \rightarrow d + \gamma \tag{2.1}$$

---

[2]The theory of the Big Bang established about years ago, is largely accepted at present as the theory of the origin of the Universe.

In a very hot environment the reverse reaction occurs so quickly that there is no appreciable accumulation of deuterium nuclei since the photon density is about 10^9 times higher than the density of neutrons or protons. However the photon energy required for occurrence of photo dissociation of deuterium is 2.226 MeV, a quantity equal to the binding energy of deuteron nucleus. The number distribution of photons in terms of energy can easily be derived from the blackbody radiation formula using the standard expression of energy density u(E) in the way

$$n(E)dE = \frac{u(E)}{E}dE = \frac{8\pi E^2}{(hc)^3}\frac{1}{e^{E/kT}-1}dE \qquad (2.2)$$

The photons have a blackbody spectrum given by Eq. (2.2) extending to very large energy E. When the number of photons in the high energy tail above the energy 2.226MeV, is less than the number of nucleons participating in the formation of deuterium, there will be very few photons to inhibit deuterium production. And the temperature for occurrence of this situation can be obtained by approximating the tail to an exponential

$$n(E)dE = \frac{8\pi E^2}{(hc)^3}e^{-E/kT}dE \qquad (2.3)$$

Integrating the above for energies $E > E_0$ we have

$$N_\gamma(E > E_0) = 8\pi\left(\frac{kT}{hc}\right)^3 e^{-\frac{E_0}{kT}}\left[\left(\frac{E_0}{kT}\right)^2 + 2\left(\frac{E_0}{kT}\right) + 2\right] \qquad (2.4)$$

Dividing this by the total number density we get the fraction F_γ for energy $E > E_0$

$$F_\gamma(E > E_0) = 0.42e^{-\frac{E_0}{kT}}\left[\left(\frac{E_0}{kT}\right)^2 + 2\left(\frac{E_0}{kT}\right) + 2\right] \qquad (2.5)$$

A plot of f in the range $10^{-11} < f < 10^{-7}$ against the corresponding E_0/kT range $21 < E_0/kT < 31$ is shown in Figure 2.2. The number of nucleons required for the formation of deuterium is obtained by the number of neutrons because of their less abundance in comparison of proton abundance. The ratio Nn/Np decreases exponentially with decreasing temperature following the relation

$$\frac{N_n}{N_p} = e^{-\frac{(m_n-m_p)c^2}{kT}} \qquad (2.6)$$

only as long as $e^\pm$ are sufficiently large and react quickly for $n \leftrightarrow p$ conversion to take place. At some temperature $T = T_f$, the ratio N_n/N_p becomes

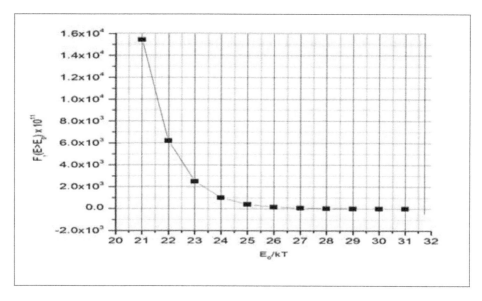

Figure 2.2: Plot of F_γ against E_0/kT in the range $21 < E_0/kT < 31$ using Eq. (2.5).

"frozen" when the rate of weak interactions becomes too small. On the basis of known cross sections for weak interactions, one can estimate this temperature as $T_f = 9 \times 10^9 K$, corresponding to $N_n/N_p \simeq 0.2$; this occurs at a time of about 3s. Neutrons, therefore originally constitutes 20% of the total number of nucleons. For nucleon to photon ratio of 10^{-9}, the critical fraction of high-energy photons required to prevent deuterium formation may be found to be 0.2×10^{-9} at $T = 9 \times 10^9 K$ at $t \simeq 250s$, which is very insensitive to the value of f and thus to the ratio N_n/N_p. Once sufficient deuterium has been formed, other nuclear reactions become possible, and there can be mass-3 nuclei:

$$^2H + n \rightarrow^3 H + \gamma \tag{2.7}$$
$$^2H + p \rightarrow^3 He + \gamma \tag{2.8}$$

Or, by

$$^2H +^2 H \rightarrow^3 H + p \tag{2.9}$$
$$^2H +^2 H \rightarrow^3 He + n \tag{2.10}$$

Resulting in the formation of ^4He:

$$^3H + p \rightarrow^4 He + \gamma \tag{2.11}$$
$$^3He + n \rightarrow^4 He + \gamma \tag{2.12}$$

The binding energies of all these reaction products are greater than the binding energy of deuterium. Thus if the photons are cool enough to permit deuterium formation, they must also permit remaining reactions to proceed. As there is no stable A=5 nuclei, ^{4}He is necessarily the end product of the above process. Again since ^{8}Be is not stable, two ^{4}He cannot combine directly, they could only combine by the intermediate production of A=7 nuclei.

$$^4He +^3 H \rightarrow^7 Li + \gamma \qquad\qquad (2.13)$$
$$^4He +^3 He \rightarrow^7 Be + \gamma \qquad\qquad (2.14)$$

In the equilibrium at the temperature $T = 9 \times 10^8 K$, the average kinetic energy for these nuclei is less than 0.1MeV which is much bellow the Coulomb barriers for these reactions. Hence all of the neutrons end as part of 4-helium having a relative abundance $N_{He}/N_p \simeq 0.081$ as calculated from frozen N_n/N_p ratio after correction for radioactive γ-decays of the neutrons between t=3s to t=250s. The relative primordial abundance of ^{4}He by weight $Y_p \simeq 0.24$ which for no additional burning of H and He expected to remain constant in the universe from t=250 s until today. The observed abundance of ^{4}He by weight $Y_p = 0.24\pm0.01$ is based on observations from a variety of astronomical systems, including gaseous nebulae, planetary nebulae, and stars including the Sun. Figure 2.3 shows the dependence of Y_p on the number of mass less species of neutrinos (at least three according to standard model) and on the nucleon to photon ratio. There will also be small concentration of primordial $^2H,^3He$, and ^7Li in the present universe. The determination of nucleon to photon ratio for deuterium is very critical as at large nucleon abundances as it is cooked into heavy nuclei very rapidly thereby reducing its own concentration. However the ratio N_d/N_p can be deduced from the shift in the absorption spectrum of atomic hydrogen caused by heavier nuclear mass of ^2H. The observed value is subject to uncertainties arising from the destruction of primordial ^2H in the evolution of galaxies and the current accepted value of $N_d/N_p \simeq 1-3\times10^{-5}$. The isotope ^{3}He is like ^{2}H a result of incomplete primordial processes and the abundance of ^{3}He decreases as the density of primordial nucleon density increases. The present-day abundances may not represent the primordial values as new ^{3}He can be produced very likely from deuterium. Thus the present value of ^{3}He abundances may be due to the primordial combination $(^3He+^2H)$ abundance. Solar observation based data suggests $(N_{2H} + N_{3He})/N_p < 6 \times 10^{-5}$. Figure 2.4 demonstrates the abundances calculated with standard model (three kinds of mass-less neutrinos). It is quite evident that the deuterium and 3He abundances constrain the nucleon to photon ratio to be greater than about 4×10^{-10} and

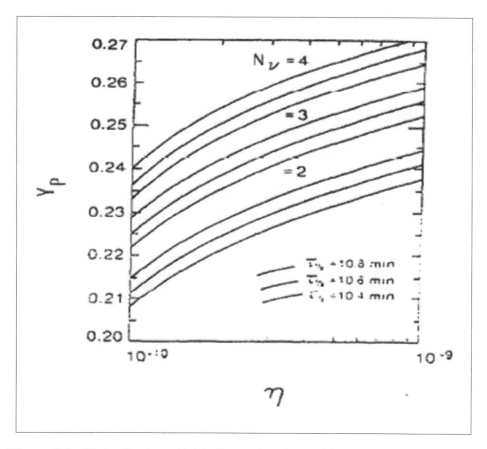

Figure 2.3: Plot of primordial helium abundance Y_p against nucleon abundance η for 2, 3 and 4 neutrinos N_ν: in each case the three curves show the range corresponding to the uncertainty in the experimental value of the neutron half-life $\tau_{1/2}=10.6$ min, $10.6 \pm 0.2min$. Ref. Schramm, D. N. and Steigman, G. Phys. Lett. B 141, 337(1984).

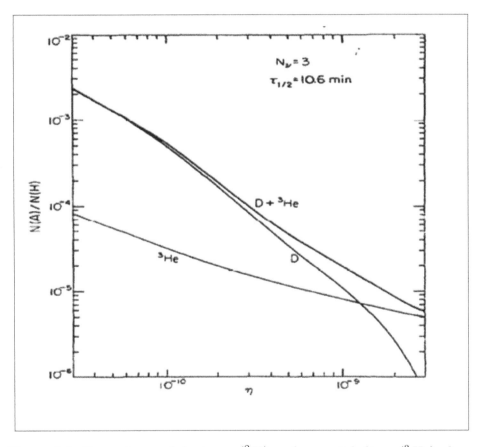

Figure 2.4: Dependence of deuteron (^2H) and mass-3 helium (^3He) abundance on the nucleon abundance. Ref: Yang, J. et al. Astrophysics J. 281, 493(1984).

Table 2.1: Reactions of proton- proton cycle and corresponding Q-values.

Reaction Stage	Reaction	Q-value (MeV)
Common to all chain:	$p + p \rightarrow d + e^+ + \nu_e$	1.442
	$d + p \rightarrow^3 He + \gamma$	5.493
PPI-chain:	$^3He +^3 He \rightarrow^4 He + p + p$	12.859
PPII-chain:	$^3He +^4 He \rightarrow^7 Be + \gamma$	1.58]
	$^7Be + e^- \rightarrow^7 Li + \gamma$	0.861
	$^7Li + p \rightarrow^4 He +^4 He$	17.347
PPIII-chain:	$^7Be + p \rightarrow^8 B + \gamma$	0.135
	$^8B \rightarrow^4 He +^4 He + e^+ + \nu_e$	18.074

from Figure 2.3, we see that this value is inconsistent with a fourth neutrino. Although much effort, both theoretical and experimental still remains undone to strengthen these conclusions, it appears that these cosmological arguments may indicate that there are no new fundamental particles beyond the present three generations of leptons and quarks.

2.3 Stellar nucleosynthesis up to iron ($A \leq 60$)

The leading process in the formation of elements of mass number $A \leq 60$ is charged particle reactions, primarily induced by protons and alpha particles. The probability of occurrence of these reactions depend on the overlapping of thermal energy distribution of particles and the Coulomb barrier penetration probability, exactly as in the case of nuclear fusion reactions. Star life begins with mixture of hydrogen and helium gas cloud. As the initial gas cloud collapses, individual atoms gains kinetic energy at the cost of their gravitational potential energy resulting in the temperature increase of the cloud. At very high temperature protons achieve sufficient kinetic energy to overcome the repulsive Coulomb barrier and fusion reactions begin. The outward radiation pressure prevents gravitational collapse and the star attains equilibrium phase like our Sun which may last for about 10^{10} years. The basic reactions of proton- proton cycle with corresponding Q-values are summarized in the table below: When a stars falls under fuel crisis due depletion of hydrogen fuel, gravitational collapse begins again and eventually higher temperature $\sim 1 - 2$ hundred million Kelvin is reached (compared to 10 million Kelvin of for the present Sun) where $\alpha - \alpha$ fission Coulomb bar-

rier can be overcome. The increased cooking temperature results in larger
radiation pressure, which in turn expands the outer envelope of the stellar
surface by a considerably large factor between 100 to 1000. The apparent
energy density of the surfaces decreases, thereby decreasing the surface temperature giving rise to the red giant stage. Since there is no stable nucleus
of mass number A=8, there are no observed end products of the reaction

$$^{4}He +^{4} He \rightarrow^{8} Be \qquad (2.15)$$

as ^{8}Be dissociates back into two 4He very quickly (with half life $T_{1/2} = 6.7 \times 10^{-17}s$) after its formation. The Q-value of the above reaction is 91.9keV
and the mean thermal energy at $2 \times 10^8 K$ is 17keV. At this condition there
will be a considerable number of high energy ^{4}He-nuclei in the tail of the
thermal energy distribution to form 8Be. A small equilibrium concentration
of the order of Boltzmann factor $\exp(-91.9kev/17keV) = 4 \times 10^{-3}$ will be
observed and the reaction rate can be determined following prescription for
D-D or D-T reactions. The large abundance of ^{12}C in the universe is well
known fact, but the calculated reaction rates of $^{4}He +^{4} He \rightarrow^{8} Be$ and
$^{8}Be +^{4} He \rightarrow^{12} C + \gamma$ reactions are not sufficient for the observed large
abundance of ^{12}C, first noticed by Fred Hoyle near the beginning of 1950s.
The Q-value of $^{8}Be +^{4} He \rightarrow^{12} C$ reaction is 7.45MeV, and Hoyle thought
that the large production of 12C demands this reaction to occur very rapidly,
which is possible if there is considerable increase in the reaction cross-section
at this energy. This prompted Hoyle to take account of resonance and he
exchanged this idea with W. A. Fowler of Cal Tech whose research group
started extended research on nuclear reactions of astrophysical importance.
The Cal Tech group discovered the ^{12}C resonance state at 7.66Mev, just
above the energy predicted by Hoyle, but well within the capability of being
reached at temperature range $1 - 2 \times 10^8 K$. The disposable Q-value of the
reaction $3\alpha \rightarrow^{12} C$ is 285keV. Afterwards Fowler et al were able to populate
the 7.65MeV excited state in ^{12}C following negative β-decay of ^{12}B. They
identified it as 0^+ state and observed its decay into 3α -particles, suggesting
its possibility of being formed by 3α's following the triple-α process

$$^{4}He +^{4} He +^{4} He \rightarrow^{12} C + \gamma \qquad (2.16)$$

The assignment of the 0^+ level is consistent with the probable s-state cou-
plings in these low-energy reactions as shown in Figure 2.5. Once ^{12}C is
formed, other α-induced reactions as tabulated below become possible: As
the Coulomb barrier (E_B) increases for heavier nuclei, the possibility of further reactions in the sequence becomes less possible. In Figure 2.6 below,

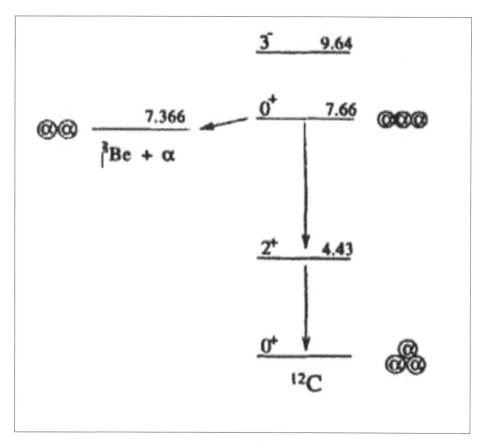

Figure 2.5: Low-lying energy levels of ^{12}C and the triple-α process. Ref: Data from C.M. Lederer and V.S. Shirley, editors. Table of Isotopes, seventh edition. Wiley, New York, 1978.

Table 2.2: Alpha particle (α) induced reactions with their Q-values and their corresponding Coulomb barrier.

Reaction	Q-value (MeV)	Coulomb barrier, EB (MeV)
$^{12}C +^4 He \rightarrow^{16} O + \gamma$	7.16	3.57
$^{16}O +^4 He \rightarrow^{20} Ne + \gamma$	4.73	4.47
$^{20}Ne +^4 He \rightarrow^{24} Mg + \gamma$	9.31	5.36

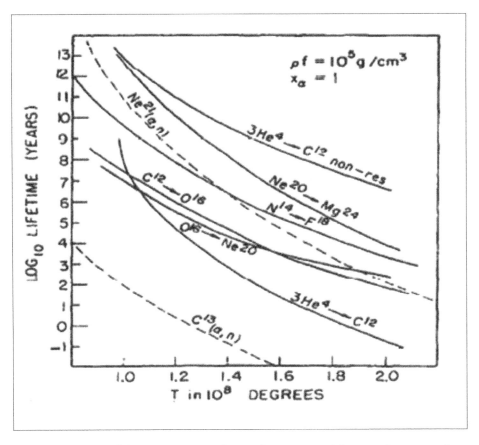

Figure 2.6: Mean lifetime in years for various α-particle reactions as a function of temperature. The reaction rate is the inverse of the lifetime, so, reactions near the bottom of the graph are the ones with the highest rates. Ref: Burbidge, E. M.; Burbidge, G. R.; Fowler, W. A. & Hoyle. F. "Synthesis of the Elements in Stars". Reviews of Modern Physics 29, 4 (1957) 547.

the calculated mean life in log scale for nuclei participating in these reactions adopted from the pioneering works of Burbidge, Burbidge, Fowler and Hoyle has been plotted against the temperature T in multiples of $10^8 K$. The striking fact to be noted here is that the ^{12}C resonance increases the reaction rate by about 8 orders of magnitude. When the helium fuel begins to be depleted, gravitational collapse sets into action again subject to the condition that the star mass is sufficiently high enough to overcome the outward radiation pressure caused by the degenerate electrons which are not ready to allow overlapping of wave functions. The star then heats up enough for burning of ^{12}C and ^{16}O through the following reactions

$$^{12}C + {}^{12}C \rightarrow {}^{20}Ne + {}^{4}He \, or ({}^{23}Na + p) \tag{2.17}$$

$$^{16}O + {}^{16}O \rightarrow {}^{28}Si + {}^{4}He \, or ({}^{31}P + p) \tag{2.18}$$

at temperatures $\sim 10^9 K$, where the Coulomb barrier can be more easily penetrated. In addition to the above reactions, other α -particle and neutron capture reactions can occur as well. For example, ^{14}N may be present in the second-generation stars, formed originally from ^{12}C as part of the carbon cycle of proton-proton fusion discussed earlier. α-capture reaction can produce the chain $^{14}N \rightarrow {}^{18}O \rightarrow {}^{22}Ne \rightarrow \rightarrow {}^{26}Mg$ ...... Reactions other than (α, γ), including (α, n) or (p, γ) will occur, with somewhat less probability. The final stage in the production of nuclei near mass-60 is silicon burning, which proceeds' in a complex sequence of reactions that happens rapidly but under quasi-equilibrium condition well in the interior of the stars. The Coulomb barrier is too high to allow direct reaction to occur instead of $^{28}Si + {}^{28}Si \rightarrow {}^{56}Ni$. The actual reaction processes that occur are combinations of photo-dissociation reactions $(\gamma, \alpha), (\gamma, p), or (\gamma, n)$ followed by capture of the dissociated nucleons:

$$^{28}Si + \gamma \rightarrow {}^{24}Mg + \alpha \tag{2.19}$$

$$^{28}Si + \alpha \rightarrow {}^{32}S + \gamma \tag{2.20}$$

together with many other analogous reactions. In the equilibrium process the Si that produced due to the burning of oxygen partially "melted" into lighter nuclei and partly "cooked" into heavier nuclei. The end products of chains of such reactions are the mass-56 nuclei ($^{56}Ni, {}^{56}Co, {}^{56}Fe$). At this point there is no longer release of energy by the capture process and the process is paused. Confirmation of this picture can be verified with reference to Figure 2.7, the abundances of elements formed by α-capture (Z=even) are far greater (an order of magnitude or more) than the neighboring odd-Z elements. For the understanding of the rate of charged-particle reactions in

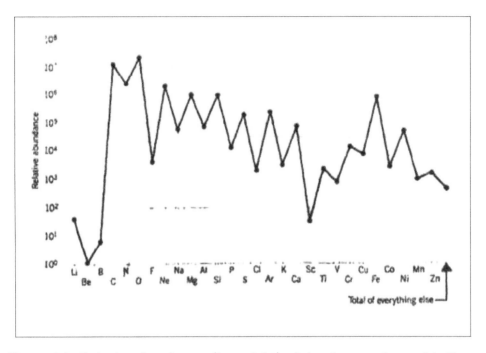

Figure 2.7: Relative abundances (by weight) of the elements beyond helium.

stellar hearts, we must try to replicate the reaction requirements using earth-based accelerators. The required energies are not large ($\sim Mev$), but we need beam of highest possible intensity (because these charged-particle reactions are strongly inhibited at low energy by Coulomb barrier penetration factors) and best possible energy resolution (to read the behavior near the discrete resonances or specific excited states). The reaction probability in a stellar environment can be estimated in way similar to the fusion reaction rate already explained in the opening chapter. The reacting particles (a + X) are described by a thermal distribution

$$n(E)dE \propto e^{-E/kT}\sqrt{E}dE \tag{2.21}$$

while the cross-section has the basic form of

$$\sigma(E) \propto (1/E)e^{-2G} \tag{2.22}$$

Where G is the Gamow factor defined as

$$G \cong (\frac{e^2}{4\pi\epsilon_0})(\pi Z_a Z_X)/h\upsilon \tag{2.23}$$

$$2G \cong (Z_a Z_X)\sqrt{A_{eff}}E^{-1/2} \tag{2.24}$$

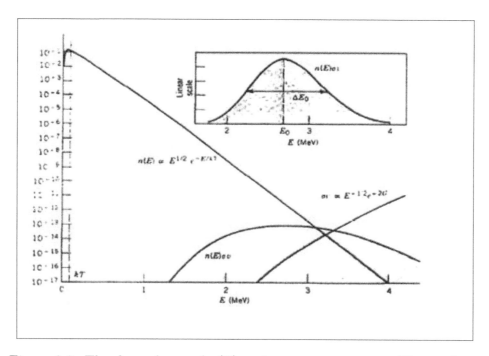

Figure 2.8: The dependence of n(E) and $< \sigma v >$ on energy. The product, which is proportional to the reaction rate is the shaded region. The inset at the top shows the reaction rate plotted on a linear scale. Illustrating the peak energy E_0 and the width ΔE_0. It is to be noted that the reaction rate peaks at energies far above kT. The curves are drawn for the reaction $^{12}C + ^{12}C$ at a temperature corresponding to kT=0.1 MeV.

All energies and velocities are given in the CM coordinates and $A_{eff} = A_a A_X / (A_a + A_X)$, the unit of energy E is MeV. The barrier penetration factor increases with increase in energy, while the number of particles decreases. Energy dependence of n(E) and $\sigma(E)$ are displayed in Figure 2.8. The reaction probability is large where graph for n(E) and $\sigma(E)$ overlap and the reactions occur at energies much larger than usual thermal energy kT. The reaction rate depends on the product σv and the number of particles available at a particular energy

$$
\begin{aligned}
Rate \quad &\propto \quad n(E)\sigma(E)v & (2.25)\\
&\propto \quad (e^{-E/kT}\sqrt{E})(E^{-1}e^{-2G})\sqrt{E} & (2.26)\\
&= \quad e^{-E/kT-2G} & (2.27)
\end{aligned}
$$

which describes the overlap region in Figure 2.9. The peak energy E_0 and

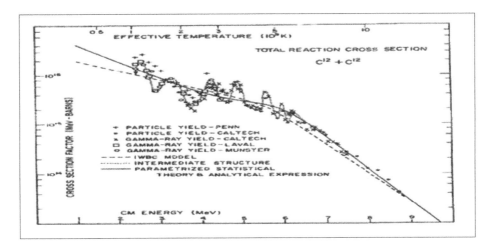

Figure 2.9: The cross-section factor S(E) for $^{12}C +^{12}C$. Ref: Fowler, W. A. Rev. Mod. Phys. 56, 149(1984).

the width ΔE_0 are given by

$$E_0 = (kTZ_aZ_X\sqrt{A_{eff}})^{2/3} \tag{2.28}$$

$$\Delta E_0 = 2^{7/6}3^{-1/2}(Z_aZ_X\sqrt{A_{eff}})^{1/3}(kT)^{5/6} \tag{2.29}$$

To obtain the information on the reaction rate at the core of a star, one should investigate reactions with particle accelerators at energy E_0, instead of kT. In CM system, the energy (E_0) required to replicate the $^{12}C +^{12}C$ reaction at stellar condition corresponding to kT=0.1MeV $(T \sim 10^9 K)$ is 2.3MeV. Rewriting eq. (9) we have

$$\sigma(E) \propto (1/E)e^{-2G}S(E) \tag{2.30}$$

Where the factor S(E) contains all information of the nuclear structure except the barrier penetration factor. For measurement of cross-section near resonance

$$S(E) = g\frac{\Gamma_{aX}\Gamma_{bY}}{(E - E_R)^2 + \Gamma^2/4} \tag{2.31}$$

Where Γ_{aX}, is the contribution of the non-coulomb factors to the entrance channel width, which involves only nuclear wave functions. The reaction rate can be computed using the value of S(E) obtained from the cross-sections:

$$S(E) = \sigma(E)Ee^{2G} \tag{2.32}$$

Figure 2.9, shows the cross-section for $^{12}C + ^{12}C$ which indicates smooth non-resonant background plus several peaks depicting some resonance structure. Since the Q- value of the $^{12}C + ^{12}C \rightarrow ^{24}$Mg has a large value of 13.9 MeV, the structure corresponds to very highly excited states of the compound nucleus. In Figure 2.10, the cross-section factor S(E) (MeV-barn) is plotted against the centre of mass energy $E_{CM}(MeV)$ for $^{12}C + ^4He \rightarrow ^{16}O + \gamma$, which shows that the structure is dominated by a single broad resonance corresponding to the alpha-unstable state at energy of about 2.5 MeV above $^{12}C + ^4He$ threshold. The calculated cross-section is in good agreement with the measured values, and the interference of two excited states of ^{16}O just below the $^{12}C + ^4$He threshold (as shown in Figure 2.11) at 7.117 MeV and 6.197MeV have to be taken into account. For the calculation of reaction rates and stellar core temperature for models of stellar structure, it is essential to accurately determine the charged-particle reaction cross-sections for energies in the MeV range in addition to the most complete knowledge about the structure and properties of the excited states. For the measurement of above types of reaction cross-sections Fowler W. A. was awarded Nobel Prize in 1983.

2.4   Stellar nucleosynthesis beyond iron(A>60)

To produce elements heavier than Fe (iron), enormous amounts of energy are needed which is thought to derive solely from the cataclysmic explosions of supernovae. A supernova is an explosion of a massive super giant star. It may shine with the brightness of 10 billion suns! The total energy output may be 10^{44}Watt-sec, as much as the total output of the sun during its 10 billion year lifetime. The likely scenario is that fusion proceeds to build up a core of iron. **The "iron group" of elements around mass number A=60 are the most tightly bound nuclei, so no more energy can be obtained from nuclear fusion.** In the supernova explosion, a large flux of energetic neutrons is produced and nuclei bombarded by these neutrons build up mass one unit at a time (neutron capture) producing heavy nuclei. The layers containing the heavy elements can then be blown off by the explosion to provide the raw material of heavy elements in distant hydrogen clouds where new stars form. From the plot of average binding energy per nucleon (B/A) versus mass number (A) graph (Figure 1.4), it is seen that the nuclear fusion reactions are not energetically favored for A>60. However for these nuclei neutron capture cross-section is large enough to govern the primary production mechanism. The most abundant element iron (Fe) has highest

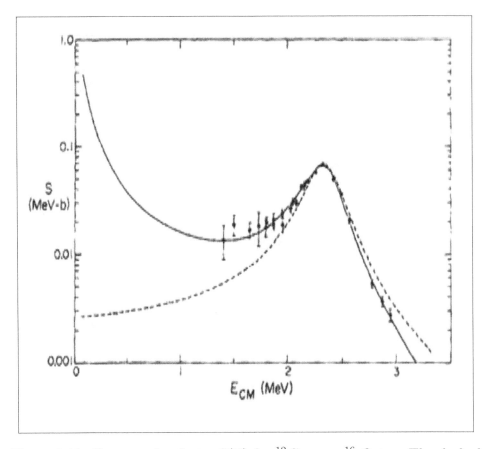

Figure 2.10: Cross-section factor S(E) for $^{12}C + \alpha \rightarrow ^{16}O + \gamma$. The dashed curve is a theoretical fit that ignores the contributions of the ^{16}O bound states, while the solid curve is a fit that includes the effect of the bound states and gives much better agreement with the experimental data. Ref: Koonin, S. E.; Tombrello, T. A. and Fox, G. Nucl Phys. A 220, 221(1974).

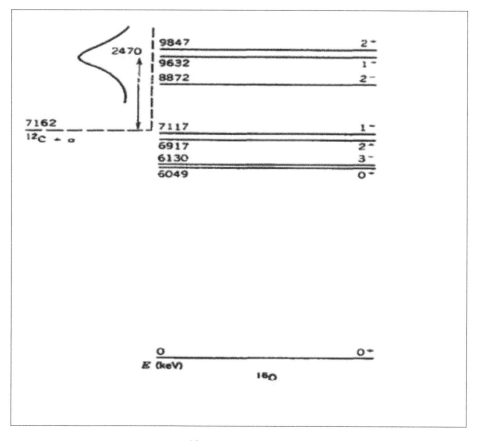

Figure 2.11: Excited states of ^{16}O. The broad resonance (which is also shown in Figure 2.8) at 2.470 MeV above $^{12}C + \alpha$ threshold includes the 1- and 2+ states at 9.632 and 9.847MeV. The interference of the 1- and 2+ states just below the $^{12}C + \alpha$ threshold (7.117 and 6.017 MeV) has a substantial effect on the calculated cross-section.

abundance of its isotope with mass-56 formed near the end of the fusion
chain reactions induced by neutron captures in a flux of neutrons:

$$^{56}Fe + n \rightarrow {}^{57}Fe + \gamma \tag{2.33}$$
$$^{57}Fe + n \rightarrow {}^{58}Fe + \gamma \tag{2.34}$$
$$^{58}Fe + n \rightarrow {}^{59}Fe + \gamma \tag{2.35}$$

The proceeding step in the process depends on the intensity of the neutron
flux. The end product ^{59}Fe is radioactive having a half-life of 45 days.
Thus if the intensity of neutron flux be too weak to produce one neutron
capture per 45 days, ^{59}Fe will produce stable ^{59}Co though β-decay. The
resulting ^{59}Co can produce radio-active ^{60}Co by capturing neutron. If the
neutron capture cross-section be large enough that the average time for
capture of neutron is small as fraction of a second, the sequence of neutron-
capture reactions can continue to ^{60}Fe ($t_{1/2} = 3 \times 10^5 yr$), ^{61}Fe ($t_{1/2} =
6min$), ^{60}Fe ($t_{1/2} = 28s$) and so on. The process continues until isotopes
with sufficiently large neutron number having half-life much shorter than the
mean-life before neutron capture is reached, and the resulting isotope then
undergo β-decay to an isotope having atomic number higher by one unit.
The sequence of neutron capture will start again until an extremely unstable
isotope of the new sequence is produced, which then undergoes to β-decay
producing isotope of element having atomic number higher by one unit than
the preceding one. The above two processes play the most important role
in the formation of the vast majority of the stable nuclear isotopes beyond
A= 60. The first process in which the neutron capture occurs only on a
long time scale, leaving enough time for all intervening β-decays to occur is
called the s-process (or slow process). The second process which does not
allow time for any decay except for the most short-lived decays is called the
r-process (or rapid process). In Figure 2.12, s- and r-process near ^{56}Fe are
shown.

It would be useful to discuss the source of the neutrons before going into
detail of neutron-capture process. Neutron emission following α-particle
induced reactions will be possible only if the target nucleus already has
loosely bound neutron(s). The neutron separation energies of α-particle
nuclei tend to be quite large in the stellar environment as shown in Table
2.3 in few cases. The most prominent reactions that occur in the stellar
environment are those with weakly bound neutrons like

$$^{13}C(\alpha, n)^{16}O \Rightarrow {}^{13}C + \alpha \rightarrow {}^{16}O + n; [Q = 2.2MeV] \tag{2.36}$$
$$^{22}Ne(\alpha, n)^{25}Mg \Rightarrow {}^{22}Ne + \alpha \rightarrow {}^{25}Mg + n; [Q = -0.48MeV] \tag{2.37}$$

Table 2.3: Neutron separation energy α-particle nuclei in few cases.

Nucleus	Neutron separation energy S_n (MeV)
^{12}C	18.7
^{16}O	15.7
^{20}Ne	16.9
^{24}Mg	16.5
^{28}Si	17.2

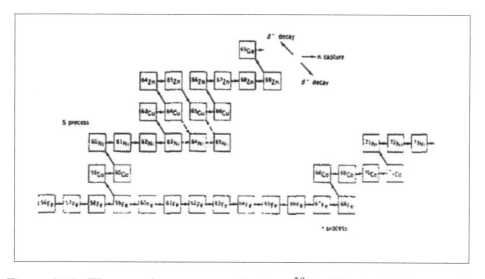

Figure 2.12: The r- and s-process paths from ^{56}Fe. The dashed lines in the s-process paths represent possible alternative routes to ^{65}Cu. Out of several other possible r-process paths due to the β-decay of the short lived nuclei, only one is shown

The above reactions will occur in the helium burning stage or red-giant phase of stellar evolution for which reaction rates can be estimated following method of previous section. The neutron density nn in the red-giant temperature range $100 - 200MK$ is about $10^{14}m^{-3}$. At this environment the reaction rate per target atom can be estimated as

$$r \simeq n_n < \sigma v > \qquad (2.38)$$

and at the upper limit of above temperature range, the speed of thermalized neutron is about $2 \times 10^{6}ms^{-1}$. A typical neutron-capture cross-section at energy $(\sim 0.002 MeV)$ is about 0.1 b. Now,

$$r \simeq (10^{14}m^{-3})(2 \times 10^{6}ms^{-1})(10^{-29}m^{2}) \quad = \quad 2 \times 10^{-9}s^{-1} \qquad (2.39)$$

or about one per 20yr which indicates an s-process situation. For r-process we need reaction rate at least 10 order of magnitude higher than the above reaction rate which can only be accomplished by large neutron flux. This is thought to be happened in the violent stellar explosion known as supernovae, but this hypothesis yet not have sound theoretical background. However there is strong experimental evidence in favor of the production of r-process nuclei. Hence we may accept that there exist site of supernovae, neutron star etc capable of producing necessary neutron flux. In the production of s-process nuclei, which occurs over a very long interval of time, we expect the establishment of an approximate equilibrium situation. Or a state of affairs in which each species has sufficient time to achieve equilibrium abundance due to equal rate of production and destruction reactions. If N_A denotes the abundance of species A and σ_A its neutron capture cross-section, then the time derivative of N_A is given by

$$\frac{dNA}{dt} = C(N_{A-1}\sigma_{A-1} - N_A\sigma_A) \qquad (2.40)$$

where C is an arbitrary constant. A is produced due to capture of neutron by nucleus of mass number A-1, and N_A is reduced due to capture of neutron by nuclei of mass number A leading to the production of nuclei of mass number A+1. At equilibrium, $\frac{dNA}{dt} = 0$, and we have

$$N_{A-1}\sigma_{A-1} = N_A\sigma_A = constant \qquad (2.41)$$

In Figure 2.13, σN has been plotted against mass number A for the nuclei beyond iron (I.e. A > 60). Above equilibrium condition is found to be violated, just above the peak in the abundance corresponding to iron since

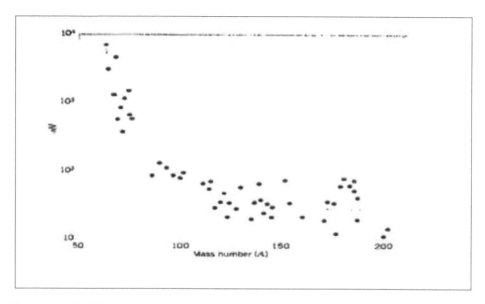

Figure 2.13: The plot of the product σN against mass number (A) shows that it approaches a constant value well above the iron (Fe) peak.

iron (Fe) abundance is not obtained through s-process. The product σN starts with relatively large values around A=60 and then decreases to the equilibrium value above A=100 having a smooth variation over a wide range (100<A<200), thereby justifying the basic assumptions regarding validity of s-process behavior. In Figure 2.14, s- and r-process paths for a small section of nuclides are shown. Some nuclei are accessible to both the s- and process paths; accounting for the abundances of these isotopes which indicates that one may be able to separate the two contributions. In the site of the gap of the isotopic sequence (between ^{120}Sn and ^{122}Sn, or between ^{121}Sb and ^{123}Sb), the s-process stops to continue along the sequence and proceeds only after undergoing β-decay to the next higher atomic number. Thus ^{122}Sn and ^{124}Sn can only be produced in the r-process. Their abundances are roughly equal, 4.5% and 5.6%, but much smaller than that of ^{120}Sn which has abundance 32.4%. An estimate by a preliminary guess reveals the fact that out of total 32.4% abundance of ^{120}Sn, 5% abundance is the contribution of the r-process and the remaining 27.4% is the contribution of the s-process. On the other hand, the β–decays at mass 122, 123, and 124 from r-process nuclei terminate at stable ^{122}Sn, ^{123}Sb, and ^{124}Sn are therefore unable to reach the tellurium (Te) isotopes of mass 122, 123, 124 respectively because of being screened from the r-process. They can only be produced by the s-process.

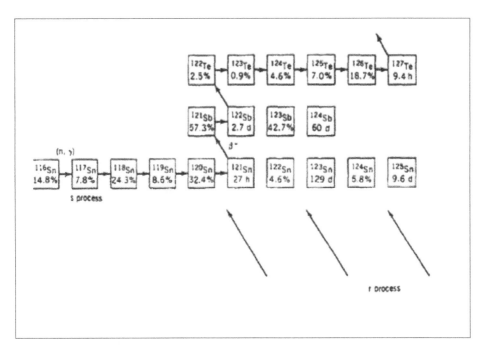

Figure 2.14: r- and s-process paths leading to Sn, Sb and Te isotopes.

Figure 2.15 shows the complete r- and s-process paths leading to the stable isotopes in the chart of the nuclide. The s-process advances in the zigzag fashion through most of the stable isotopes just above A=209 through which the s-process could proceed. The r-process has no such restriction and can continue until the fission half-lives are as short as the r-process capture times. Near the finishing end of the r-process, it may be possible to produce massive nuclei; such a possibility has inspired the search for evidence for super heavy nuclei in natural materials, so far without breakthrough. Near the magic numbers the β–decay times becomes so small that an added neutron decays to a proton in a smaller time compared with the r-process capture time. This accounts for the upward movement of the r-process path at N=50, 82, and 126. As these nuclei subsequently β–decay towards the stable isobars, a slight overabundance of stable nuclei will emerge. This will occur near A=80, 130 and 195, as shown in Figure 2.16.

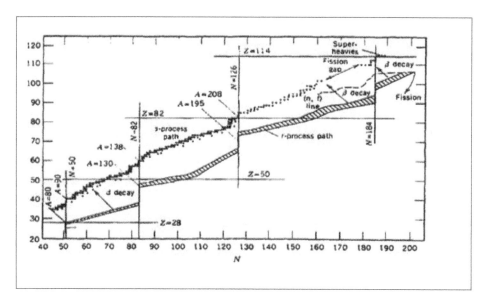

Figure 2.15: Neutron capture paths for r and s- processes.

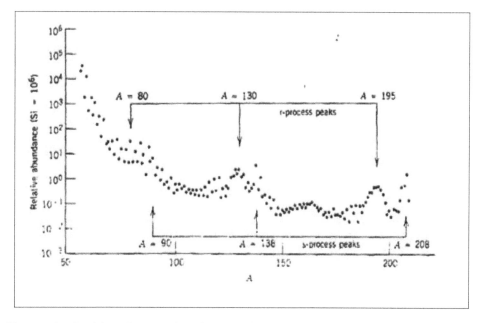

Figure 2.16: Abundance of isobars. The peaks around A=80, 30 and 195 originate from the beta-decays of r-process progenitors with N=50, 82, or 126. The peaks around A=90, 138, and 208 result from s-process stable nuclei with N=50, 82, or 126. The difference between abundances of odd-A and even-A are notable.

Solved Problems

1. List the nuclides that play active role in the bing bang nucleosynthesiss (BBN).

   Solution There are 26 nuclides that took active participation in the BB nucleosynthesis processes. These are- n, p, ^{2}H, ^{3}H, ^{3}He, ^{4}He, ^{6}Li, ^{7}Li, ^{7}Be, ^{8}Li, ^{8}B, ^{9}Be, ^{10}B, ^{11}B, ^{11}C, ^{12}B, ^{12}C, ^{12}N, ^{13}C, ^{13}N, ^{14}C, ^{14}N, ^{14}O, ^{15}N, ^{15}O, and ^{16}O.

2. Define Q-value of a nuclear reaction. How can you calculate Q-value of a nuclear reaction in terms of mass excess of the participating nuclides? Give example.

   Solution The Q-value of a reaction of a typical nuclear reaction of the type $A+B \rightarrow C+D$ is defined as the difference of the rest mass energy of the nuclides in the entrance channel to that in the exit channel $Q/c^2 = \{(M_A + M_B) - (M_C + M_D)\} = \Delta(A) + \Delta(B) - \Delta(C) - \Delta(D)$ where mass excess[3] $\Delta(X) = [M(X) - A_X m_u]c^2$, M(X) is the atomic mass of nuclide X, A_X its mass number, $m_u = 931.5 MeV/c^2 \simeq 1.66 \times 10^{-27} kg$ is the atomic mass unit $= (1/12)$ of the mass of a ^{12}C atom.

   Example For the reaction $^6Li +^2 H \rightarrow \alpha\alpha$, $\Delta(^6Li)$=+14.087MeV, $\Delta(^2H)$=+13.136MeV, $\Delta(^4He)$=+2.425MeV. $Q = \Delta(^6Li) + \Delta(^2H) - 2 \times \Delta(^4He) = 14.087$MeV +13.136MeV- 2(2.425 MeV)= + 22.373 MeV.

Exercise

1. The greatest binding energy per nucleon occurs near ^{56}Fe and is much less for ^{238}U. Explain this in terms of the semiempirical nuclear binding theory. State the semiempirical binding energy formula (no need to specify the values of the various parameters).

2. List the reactions in the three main branches of the proton-proton reaction chain, giving estimates for the relative branching ratios for each part of the chain.

3. Explain the term 'radiative capture reaction' as used in nuclear physics.

4. Calculate the Q value for the reaction $d + d \rightarrow \alpha$ given that he masses of the deuteron and alpha particle are M(d)=2.014102u and

---

[3]Mass excess (Δ) of nuclides are available in tabular form in the Nuclear Wallet Cards, Obtainable from J. K. Tuli, National Nuclear Data Center, BNL, Upton, New York, USA or www.nndc.bnl.gov

M(α)=4.002603u. Why is the reaction thought to play a negligible role in the production of α in the p-p chain?

5. Calculate the energy of the photon emitted in the direct capture reaction $^3He(\alpha, \gamma)^7Be$ to the excited $\frac{1}{2}^-$ state in ^{7}Be with an excitation energy of 429 keV. Assume that the ^{3}He nucleus is at rest and that the α particle has a kinetic energy of 100 keV in the laboratory frame. Note: M(^4He)=4.002603u, $M(^3He)$=3.016029u, $1u = 931.5 MeV/c^2$ and $M(^7Be)$=7.0169289u.

References

Quoted in the text:

- Eddington, A. S. (1919). "The sources of stellar energy". The Observatory 42: 371–376. .

- von Weizsäcker, C. F. (1938). "Über Elementumwandlungen in Innern der Sterne II" [Element Transformation Inside Stars, II]. Physikalische Zeitschrift 39: 633–646.

- Bethe, H. A. (1939). "Energy Production in Stars". Physical Review 55 (5): 434–456

- Chadwick, James (1932). "Possible Existence of a Neutron". Nature 129 (3252): 312.

- Suess, Hans E.; Urey, Harold C. (1956). "Abundances of the Elements". Reviews of Modern Physics 28 (1): 53.

- P.W. Merrill (1956). "Technetium in the N-Type Star ^{19}PISCIUM". Publications of the Astronomical Society of the Pacific 68: 400.

- Diehl, R.; et al. (1995). "COMPTEL observations of Galactic ^{26}Al emission". Astronomy and Astrophysics 298: 445.

- Iyudin, A. F.; et al. (1994). "COMPTEL observations of Ti-44 gamma-ray line emission from CAS A". Astronomy and Astrophysics 294: L1.

- Davis, Raymond; Harmer, Don S.; Hoffman, Kenneth C. (1968). "Search for Neutrinos from the Sun". Physical Review Letters 20 (21): 1205.

- Tang, X. D.; et al. (2007). "New Determination of the Astrophysical S Factor SE1 of the $C^{12}(\alpha,\gamma)O^{16}$ Reaction". Physical Review Letters 99 (5): 052502

Chapter 3

Sources of Energy in Stars

3.1 Introduction

Nuclear physics gives a self-consistent picture of the energy source for the Sun and its subsequent lifetime, as the age of the solar system derived from meteoritic abundances of lead and uranium isotopes is about 4.5 billion years. As the mass of the Sun has enough nuclear fuel to allow for core hydrogen burning on the main sequence of the HR-diagram via the pp-chains for about 9 billion years, a lifetime primarily set by the extremely slow production of deuterium via the thermonuclear fusion reaction,

$$^1_1H + ^1_1H \rightarrow ^2_1D + e^+ + \nu_e + 0.42 MeV \tag{3.1}$$

As stars produce their energy by nuclear fusion reactions, the positive charges of two reacting nuclei will try to oppose the nuclei getting too close together by forming a Coulomb Barrier between them. Classically, for two nuclei to fuse together, their relative kinetic energy must be greater than this Coulomb barrier as shown in Figure 3.1, R_c indicates the classical turning end for the given projectile energy. In stars, this kinetic energy of the nuclei is generated by the gravitational contraction of the stars. In its normal phase, this gravitational contraction is balanced by the existing gas pressure. The kinetic energy of the nuclei also comes from the energy excess given off in the nuclear fusion reactions which occur in the heart of the star. If two nuclei fuse together, their total mass/ energy will either be greater or less than the mass/energy of the two reacting nuclei. A particular nucleus is described in terms of its constituent number of protons (Z), neutrons (N) and the mass number (A) which is equal to the total number of nucleons (N+Z). The total mass energy of a nucleus is less than the sum of the masses of the

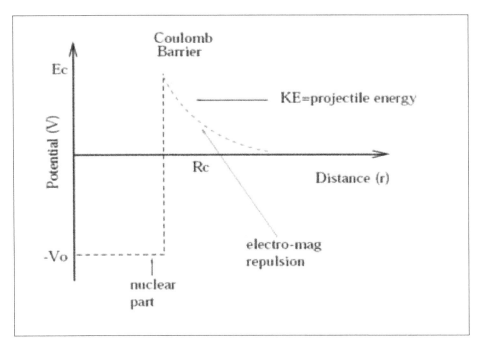

Figure 3.1: Coulomb barrier experienced by a charged particle directed to a positively charged nuclear target for nuclear reaction.

constituent protons and neutrons due to the binding energy of the system. This binding energy can thus be calculated by

$$E_B \;=\; \Delta E = \Delta M_N c^2 = (M_N - Z M_p - N M_n) c^2 \qquad (3.2)$$

in which M_N = nuclear mass, M_p = proton mass, M_n = neutron mass and c is the speed of light in free space. The quantity ΔE is thus the energy required to separate the bound nucleus into it constituent nucleons. As stated above, nuclear fusion reactions are prime source of the vast energy of shining stars. The different fusion reaction chains or cycles or processes responsible for the energy production of the stars include:

1. Hydrogen burning: p-p (proton-proton) chain;

2. Carbon-nitrogen-oxygen (CNO) and other hydrogen burning cycles;

3. Helium burning (triple-α) process;

4. Fusion of alpha conjugate nuclei.

The above fusion processes apart from energy production are also responsible for the creation of many chemical elements with mass A<56. For nuclei with A>56 it costs energy to fuse 2 nuclei together.

3.2 Hydrogen burning and p-p (proton-proton) chain

According to the theory of stellar structure and evolution stars are self-gravitating balls of (mostly) hydrogen gas that act as thermonuclear furnaces to convert the primary product of the big bang (hydrogen) into the heavier elements of the periodic table. For most of its life a star maintains a state of stable equilibrium in which the in-ward force of gravity is balanced by the outward force of pressure. This internal pressure is generated by the energy released in the nuclear reactions which burn hydrogen at the star's center. These nuclear reactions are also the source of the star's luminosity. During the time the star burns hydrogen in its core, it maintains a fixed radius and luminosity and consequently a constant surface temperature. The exact value of the equilibrium radius, luminosity and surface temperature of a star depends almost entirely on one parameter, the star's mass. The more massive the star, the greater its luminosity, size and surface temperature. Collectively, hydrogen burning stars of varying mass form a well defined locus of points on the observable luminosity-effective temperature plane called the HR diagram. This locus of points is called the main sequence and the hydrogen burning phase of a star's life is known as the main sequence phase. Main sequence stars have their masses in the range 0.08 to $100 M_\odot$ (solar masses). Stars with smaller masses have insufficient mass to raise their central temperatures enough to enable hydrogen fusion, such objects are referred to as brown dwarfs. Stars with larger masses are presumably too radiant to hold on to their outer atmospheres. In main sequence stars (such as the sun), the main energy source is thought to come from converting four protons (hydrogen nuclei) into and an α-particle ($^4_2$He nucleus) which can be achieved in a number of ways including the proton-proton chain as discussed below. The basis of the proton-proton chain is to turn four protons in a $^4_2$He nucleus together with two electrons and two neutrinos.

$$4 ^1_1 H \to ^4_2 He + 2e^+ + 2\nu_e \qquad (3.3)$$

Because the p + p di-proton (^{2}He) system is unbound, the first step in this reaction is thought to be the three body reaction,

$$^1_1 H + ^1_1 H \to ^2_1 H + e^+ + \nu_e \qquad (3.4)$$

having Q-value of 1.44 MeV. It is to be noted that this reaction proceeds under weak interaction and has a very low probability (or reaction cross-section). It relies on the uncertainty principle ($\Delta E \Delta t = \hbar$) which means that occasionally, two protons can be quasi-bound in a resonant, di-proton state for long enough for the weak interaction to occur. Note that this (arguably the most important nuclear reaction in astrophysics) has never been observed experimentally. Following the formation of the deuteron, the following, proton capture reaction is very likely to occur.

$$4 ^2_1 H + ^1_1 H \to ^3_2 He + \gamma \qquad (3.5)$$

which has a Q-value of +5.49 MeV. Note that while the reaction $d + d \to ^4 He$ is possible, it is less likely due to the relatively small number of deuteron present compared to the vast 'sea' of protons. Theoretical estimate reveals the fact that in the sun about 1 deuteron is formed for every 10^{18} protons. Once the $^3_2$He nucleus is formed, it cannot capture another proton since the compound system $^4_3$Li is unbound, i.e.

$$^3_2 He + ^1_1 H \to ^4_3 Li* \to ^3_2 He + ^1_1 H \qquad (3.6)$$

However, other possible fusion reactions involving the $^3_2$He nucleus include fusion with a second $^3_2$He nucleus to form a $^4_2$He (α-particle) and two protons,

$$^3_2 He + ^3_2 He \to ^4_2 He + 2 ^1_1 H + \gamma, [Q = 12.86 MeV] \qquad (3.7)$$

or fusing with an alpha-particle to form $^{7}_{4}$Be as below.

$$^{3}_{2}He + ^{4}_{2}He \rightarrow ^{7}_{4}Be \tag{3.8}$$

The $^{7}_{4}$Be can then be used to form two α particles using either of the following reactions,

$$^{7}_{4}Be + e^{-} \rightarrow ^{7}_{3}Li + \nu_{e} \tag{3.9}$$

$$^{7}_{3}Li + p \rightarrow ^{4}_{2}He + ^{4}_{2}He \tag{3.10}$$

or

$$^{7}_{4}Be + p \rightarrow ^{8}_{5}B \tag{3.11}$$

$$^{8}_{5}B \rightarrow ^{8}_{4}Be^{*} + e^{+} + \nu_{e} \tag{3.12}$$

Here it is to be noted that ^{8}Be is unbound.

$$^{8}_{4}Be \rightarrow ^{4}_{2}He + ^{4}_{2}He \tag{3.13}$$

Note that in all cases, some of the energy released in these reactions will be taken away by the neutrino. In all three branches of the p-p chain, the total energy released is the same (26.73 MeV), but the fraction which is carried away by the neutrinos differs. The effective energies remaining in the chains (Q_{eff}) are 26.20 MeV (chain I), 25.66 MeV (chain II) and 19.17 MeV (chain III). Note that the weak interactions,

$$^{7}_{4}Be + e^{-} \rightarrow ^{7}_{3}Li + \nu_{e} \tag{3.14}$$

$$^{8}_{5}B \rightarrow ^{8}_{4}Be + e^{+} + \nu_{e} \tag{3.15}$$

have nuclear structure effects which determine the average energy of the neutrinos (i.e. the effective energy which is lost in the process). Positron decay of $^{8}_{5}$B goes mainly to an excited (unbound) resonance state in $^{8}_{4}$Be which consequently decays into two α–particles.

3.3   CNO Cycles

Stellar interiors in main sequence stars are made up mostly of hydrogen (Z=1) and helium (Z=2). There are also small quantities of heavier elements like carbon (Z=6) which is particularly relevant in the case of more massive stars. Most of the present stars are thought to be second or third generation stars (also known as population 1 type stars) made up from hydrogen and helium mixed with some heavier nuclei formed when older stars

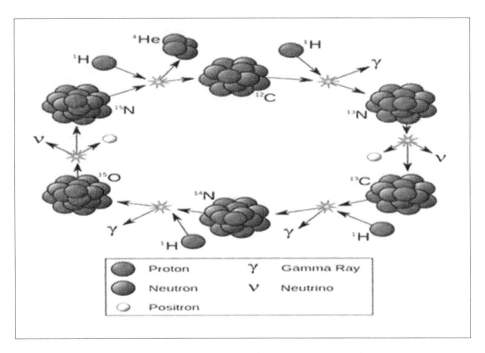

Figure 3.2: The simple CN (carbon-nitrogen) cycle.

met violent explosion termed as supernova. In these types of stars, energy can be released by burning 4 protons into a $^4_2$He nucleus (as in the p-p chains) via proton reactions on heavier elements starting with oxygen. The simplest and commonest of these reaction-cycles uses the fusion of a proton with a $^{12}_6$C nucleus to start the carbon-nitrogen (-oxygen) or CN(O) cycle.

$$^{12}_6C + ^1_1H \quad \rightarrow \quad ^{13}_7N + \gamma + 1.95 MeV \tag{3.16}$$

$$^{13}_7N \quad \rightarrow \quad ^{13}_6C + e^+ + \nu_e + 1.20 MeV (\tau_{1/2} = 9.965 min) \tag{3.17}$$

$$^{12}_6C + ^1_1H \quad \rightarrow \quad ^{14}_7N + \gamma + 7.54 MeV \tag{3.18}$$

$$^{15}_8O \quad \rightarrow \quad ^{15}_7N + e^+ + \nu_e + 1.73 MeV (\tau_{1/2} = 122.24 sec) \tag{3.19}$$

$$^{15}_7N + ^1_1H \quad \rightarrow \quad ^{12}_6C + ^4_2He + 4.96 MeV \tag{3.20}$$

Note that the initial $^{12}_6$C material is not 'used up' in the reactions and can thus be used over and over again, thus the name cycle. A simple CN cycle is shown in Figure 3.2. The net result is again four protons converted into a ^{4}He together with two positrons and two electron neutrinos, the same final result as the PP-chains. The only exception is that ^{12}C is used as the catalyst here. The simple CN cycle discussed above assumes the reaction,

$$^{15}_{7}N +^{1}_{1}H \rightarrow^{16}_{8}O^* \rightarrow^{12}_{6}C +^{4}_{2}He \tag{3.21}$$

However, there are very little chances (about one in every thousand re-actions) that the excited $^{16}_{8}O$ nucleus may decay to the ground state via γ-emission, instead of dissociating into α-particle and ^{12}C nucleus, thereby breaking out of the simple CN cycle. Thus, the latter part of the CNO cycle will contribute only a very small portion of the overall energy released in the reactions, because ^{16}O decay to its ground state only once in 1000 cycles. However, it plays a very crucial role in the nucleosynthesis (formation) of ^{16}O. There are several side chains to the main CNO cycle that are also of interest. The ^{13}N produced in the (p, γ) reaction on ^{12}C may be converted through another (p, γ) reaction into ^{14}O, which then β^+-decays to ^{14}N,

$$^{13}_{7}N +^{1}_{1}H \quad \rightarrow \quad ^{14}_{8}O + \gamma \tag{3.22}$$
$$^{14}_{8}O \rightarrow^{14}_{7}N + e^+ + \nu_e \tag{3.23}$$

The final product returns the process to the main CNO cycle in the form of ^{14}N. Similarly, some of the ^{15}N near the end of the main cycle may be converted back to ^{14}N by the following chain of reactions:

$$^{15}_{7}N +^{1}_{1}H \quad \rightarrow \quad ^{16}_{8}O + \gamma \tag{3.24}$$
$$^{16}_{8}O^{1}_{1}H \quad \rightarrow \quad ^{17}_{9}F + \gamma \tag{3.25}$$
$$^{17}_{9}F \quad \rightarrow \quad ^{17}_{8}O + e^+ + \nu_e \tag{3.26}$$
$$^{17}_{8}O +^{1}_{1}H \quad \rightarrow \quad ^{14}_{7}N +^{4}_{2}He \tag{3.27}$$

Again a ^4He nucleus is made from four protons by this procedure. However, it is important to note here that the last reaction

$$^{17}_{8}O +^{1}_{1}H \rightarrow^{4}_{2}He +^{14}_{7}N \tag{3.28}$$

returns most of the catalytic material back into the original CN cycle. The complete CNO cycle including the breakout's from the main cycle is shown in Figure 3.3. The CN cycle burns hydrogen at a much faster rate than the p-p chain, because of the delay associated with the $p+p \rightarrow d + e^+ + \nu_e$ reaction. The CNO cycle is however slowed down due to the larger Coulomb barrier associated with the protons fusing with heavier-Z nuclei such a carbon. The CNO cycle is thought to dominate over the p-p chain because of its suitability of being the major way of energy production in larger and more massive stars as they have much higher internal temperatures. This means that the

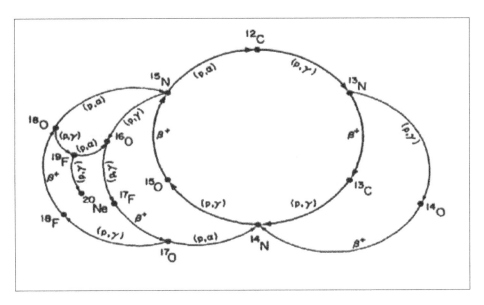

Figure 3.3: Complete carbon-nitrogen-oxygen (CNO) cycle (including break-outs from CN cycle) of nucleosynthesis showing the different reactions involved in converting protons into ^{4}He.

thermal energy of the protons is much higher and they can tunnel through the Coulomb barrier more easily than in stars having smaller sizes, smaller masses and lower temperatures. The probability of tunneling (P_t) through the Coulomb barrier is approximately given by

$$P_t = e^{-\sqrt{E_G/E}} \tag{3.29}$$

where E_G represents the Gamow energy determining the height of the Coulomb barrier,

$$E_G = 0.978(Z_A Z_B)^2 \tag{3.30}$$

From the above equation it can easily be understood that higher the energy (E) of the reacting particles higher will be the tunneling probability, thereby justifying the forgoing remarks. The ^{17}O in the intermediate step above may return some ^{15}N to the main cycle and produce a ^{4}He nucleus through the chain of reaction

$$^{17}_{8}O + ^1_1 H \quad \rightarrow \quad ^{18}_{9}F + \gamma \tag{3.31}$$

$$^{18}_{9}F \quad \rightarrow \quad ^{18}_{8}O + e^+ + \nu_e \tag{3.32}$$

$$^{18}_{8}O + ^1_1 H \quad \rightarrow \quad ^{15}_{7}N + ^4_2 He \tag{3.33}$$

The ^{18}O produced in the intermediate stage here may also undergo further proton capture and produce ^{19}F through the process

$$^{18}_{8} + ^{1}_{1}H \rightarrow ^{19}_{9}F + \gamma \tag{3.34}$$

$$^{19}_{9}F + ^{1}_{1}H \rightarrow ^{16}_{8}O + ^{4}_{2}He \tag{3.35}$$

Some of the ^{19}F, in turn, may be converted into ^{20}Ne by a further (p, γ) reaction,

$$^{18}_{8} + ^{1}_{1}H \rightarrow ^{19}_{9}F + \gamma \tag{3.36}$$

$$^{19}_{9}F + ^{1}_{1}H \rightarrow ^{20}_{10}Ne + \gamma \tag{3.37}$$

In terms of the amount of energy produced, the PP-chains still constitute the dominant source. However, the CNO cycle is able to generate some of the heavier elements that are of interest from a nucleosynthesis point of view. Note that in all of these CNO reactions, the initial catalytic material remains present in the cycle and can be re-used over and over again. The reaction rates for the different (p, α) and/or (p, γ) reactions in the cycle depend crucially on the stellar temperature (i.e. proton energy).

3.4 The Hot CNO Cycle and the Rapid Proton Process

At very high temperatures (i.e. particle energies) and particles densities, such as the conditions in very massive stars, supernova or neutron stars, the hydrogen burning will occur at very high temperatures of $10^8 \rightarrow 10^9 K$. For these conditions, in the CNO cycle, the lifetimes of some of the beta unstable systems such as ^{13}N and ^{15}O are long enough (and the proton density high enough) that proton capture can occur on these unstable nuclei before they undergo β-decay. We may refer the following reaction as an example

$$^{13}_{7}N + ^{1}_{1}H \rightarrow ^{14}_{8}O + \gamma \tag{3.38}$$

This high temperature and density limit is known as the hot or β-limited CNO cycle. The rate of burning of 4 protons to a α-particle is limited by the β-decay lifetimes of the unstable, proton rich $^{14}_{8}$O and $^{15}_{8}$O nuclei. Note that at lower temperatures, the $^{14}_{7}N(p, \gamma)^{15}_{8}$O reaction sets this limit. For $T > 5 \times 10^8 K$, the catalytic material from the hot CNO cycle can be lost or break out and form other nuclei. This breakout and the subsequent proton capture of these breakout nuclei is thought to play an important part in the

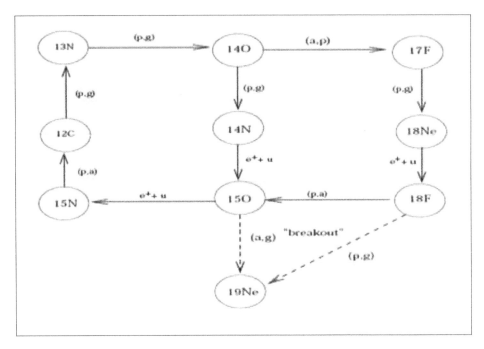

Figure 3.4: The hot or β-limited CNO cycle.

synthesis of heavier elements. The first breakout reaction from the hot CNO cycle as shown in Figure 3.4 is,

$$^{18}_{9}F +^1_1 H \rightarrow^{19}_{10} Ne + \gamma \qquad (3.39)$$

Once the material has broken out of the CNO cycle, it can rapidly capture extra protons, making more beta-unstable nuclei. This is known as the rapid-proton or rp-process (shown in Figure 3.5) and is analogous to the standard rapid-neutron or r-process for neutron capture. At each step along the rp process, the material can either take on an extra proton via proton capture or have to wait for a γ-decay if for example the extra proton will form a compound nucleus which is unbound to proton emission or the beta-lifetimes is so fast that it can compete with the rate of proton capture. Note that unlike the r process, the rate of the rp-process is hindered for heavier nuclei by the increasing Coulomb barrier experienced by the proton trying to fuse with higher Z systems. Thus for proton reactions to occur for heavier nuclei, higher and higher temperatures (and densities) are required (see Figures 3.5 & 3.6). The rp-process can also lead to the population of nuclei (for example $^{20}_{10}$Ne and $^{24}_{12}$Mg), from which heavier hyrdogen burning cycles analogous to the CNO cycle can be formed. These include the NeNa and MgAl cycles (see

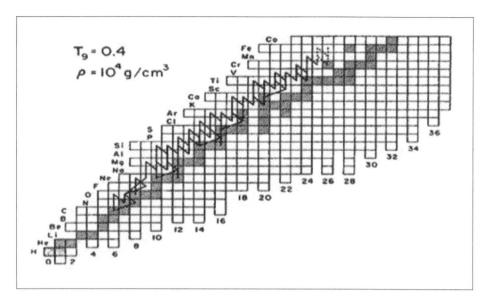

Figure 3.5: Predicted rp-process at medium T and ρ (from Champagne and Wiescher).

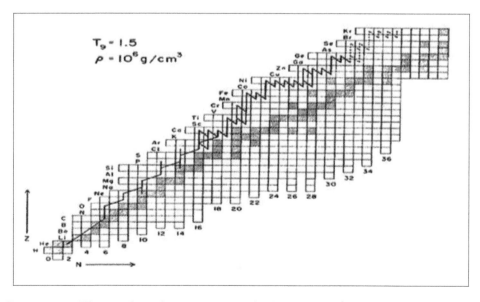

Figure 3.6: The predicted rp-process at high T and ρ (from Champagne and Wiescher).

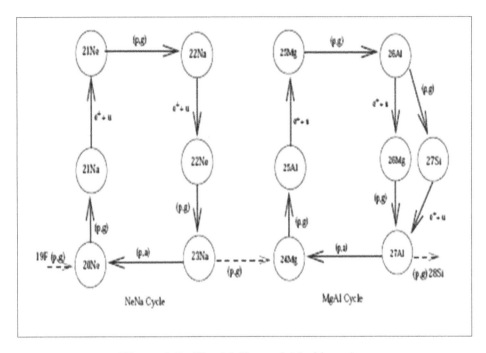

Figure 3.7: The NeNa and MgAl cycles.

Figure 3.7). Note that due to the relatively high Coulomb barriers involved in these cycles, they are relatively unimportant as energy sources in stars. The path of the rp process (i.e. which nuclei are populated) depends crucially on the temperature and density conditions. Note that at very high temperatures and densities ($T \sim 1.5 \times 10^9 K, \rho \sim 10^6 g/cm^3$), the main reactions up to Argon nuclei (Z=18) are (α, p) reactions, where an α-particle, rather than a proton is captured. These reactions are less likely at low temperatures due to the higher Coulomb barrier experienced for alpha-particles (Z=2) compared to protons (Z=1).

3.5   The Early rp–Process and Bottleneck Reactions.

The breakout of the hot CNO cycle via the $^{15}O(\alpha, \gamma)^{19}Ne$ reaction can give rise to the early stages of the rp-process, via the reaction $^{19}Ne(p, \gamma)^{20}Na$ and $^{20}Na(p, \gamma)^{21}Mg$ Proton capture reactions for nuclei with A >20 close to the line of β-stability (N = Z <20) have typical Q-values of around 5MeV. However, the reaction Q-values of the $T_z = -1/2$ nuclei ($T_z = ((N - Z)/2)$, eg $^{19}_{10}Ne$, $^{23}_{12}Mg$, $^{27}_{14}Si$, $^{31}_{16}S$ etc. are only $\leq 2MeV$. Therefore the reaction rates are determined by the presence of isolated nuclear states or resonances (and by non-resonant capture). These reactions are known as bottleneck reactions as they represent the only link to heavier elements for $T \leq 10^9 K$. Their reaction rates effectively limit the overall reaction flow in the rp process.

3.6 Helium (⁴He) Burning

For a massive star having the temperature and density high enough, after the exhaust of most of the hydrogen in the star due to transmutation into helium, the stellar core can be hot enough for helium burning to take place. Helium burning plays an important role for massive (very hot and dense) stars in both energy production and in the nucleosynthesis of heavy elements.

3.7 The Triple-α Process

Note that there are no particle stable isotopes with A=5 or 8. Thus the reactions $p + \alpha$ and $\alpha + \alpha$ form unstable systems $^5_3 Li^*$ and $^8_4 Be^*$. However, $^{12}_6 C$ is the fourth most abundant nucleus in nature. Thus one of the major questions in nuclear astrophysics is how the instability gaps at A=5 and 8 were bridged to form so much $^{12}_6 C$. The idea of a simultaneous interaction of three α-particles can be ruled out purely by the vanishingly small probability of this type of interaction compared to the large amounts of ^{12}C observed. The solution is thought to be that ^{12}C is formed in a two step process, involving resonant states in both the unbound $^8_4$Be system and above the particle decay threshold energy in the excited form of ^{12}C. In the first step of this process (known as the triple-α process), two $^4_2$He nuclei combine to form a resonant state in $^8_4$Be. As discussed earlier, $^8_4$Be is energetically unbound against breakup, back into 2α-particles (by only 92 keV). However, by the uncertainty principle, the resonant state can have a finite lifetime ($\Gamma\tau = h$), and the $^8_4$Be nucleus can live for an average of 10^{-16} seconds. This lifetime, while quite small, is large compared to the typical transit time of two α-particles with kinetic energies close to the extra energy required to populate the unbound resonant state (92 keV) of about 10^{-19s}. In this way, it is possible to build up an equilibrium concentration in the unstable $^8_4$Be, which is continually being created and decaying into two α-particles, i.e. the rate of production of $^8_4$Be is equal to the rate of break-up into two alphas. The equilibrium concentration of $^8_4$Be in a $^4_2$He environment can be

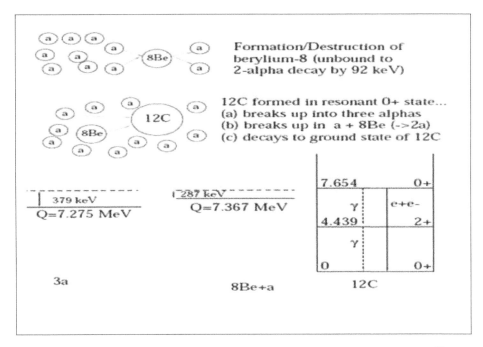

Figure 3.8: Schematic of the triple-α process in the formation of $^{12}_{6}$C.

calculated by the Saha equation, i.e

$$n_{12} = \frac{n_1 n_2}{1 + \delta_{12}} \left(\frac{2\pi}{\mu kT}\right)^{3/2} \hbar^3 \omega e^{-\frac{E_R}{kT}} \qquad (3.40)$$

where n_{12} are the number of nuclei in the resonant state ($E_R = 92 keV$ for $^8_4$Be, n_1 and n_2 are the number of nuclei of type 1 and 2 (for $2\alpha \rightarrow ^8_4$ $Be, 1 = 2$), δ_{12} =Kronecker delta, $\mu = \frac{m_1 m_2}{m_1 + m_2}$ =reduced mass of the system, k=Boltzmann's constant, E_R is the energy of the resonant state, T is the temperature and ω is a statistical factor. For $T = 10^8 K$, and a density of $10^5 g/cm^3$, the Saha equation suggests $N(^8_4 Be)/N(^4_2 He) \simeq 5 \times 10^{-10}$. As Figure 3.8 shows, the second stage of the ^{12}C formation requires the capture of a third alpha particle by the resonant state in $^8_4$Be, followed by a gamma-decay to the ground state of $^{12}_6$C, via the reaction $^8_4 Be(\alpha, \gamma)^{12}_6$C. Hoyle suggested that in order for the above reaction to proceed at a fast enough rate to account for the natural abundance of ^{12}C, there had to be resonant state at energies close to the breakup threshold energy of ^{12}C into an α-particle and a $^8_4$Be resonance. More specifically, this state should have spin/parity 0^+. The existence of such resonant state, greatly enhances the nuclear reaction rate and in experiment, such a state was observed at an excitation of 7.654

MeV in $^{12}_{6}C$ which is 287 keV above the $^{8}_{4}Be + \alpha$ breakup threshold. And hence, 287+92=379 keV above the $\alpha + \alpha + \alpha$ breakup threshold.

3.8 Heavier Element α-Burning

Once $^{12}_{6}C$ has been formed in the triple alpha-process, radiative capture (α, γ) reactions can occur to form other α-conjugate nuclei. For example $^{16}_{8}O$ can be formed via the $^{12}C(\alpha, \gamma)^{16}O$. In principle, if the stellar temperature is high enough (to tunnel through the increased Coulomb barrier), this α-capture chain can continue, allowing the formation of $^{20}_{10}Ne, ^{24}_{12}Mg, ^{28}_{14}Si \ldots$ etc. However, the rates for these reactions are highly sensitive to the nuclear structure of the individual nuclei involved, and as we shall see later, the presence of low-lying resonances plays a vital role.

3.8.1 Neutron Production in (α, n) Reactions.

In second generation stars, helium burning reactions can take place with other, non-alpha-conjugate nuclei which are present, such as ^{13}C and ^{22}Ne. These reactions produce neutrons, i.e., $^{13}_{6}C(\alpha, n)^{16}_{8}O$ and $^{22}_{10}Ne(\alpha, n)^{25}_{12}Mg$ Such reactions are thought to be the source of neutrons which play a role in the formation of heavier elements (above ^{56}Fe) via the slow neutron or s-process.

3.9 The s and r-Neutron Processes.

As the binding energy per nucleon curve shows, fusion is no longer energetically favored for compound nuclei with A>56. However, clearly, nuclear species exist above this mass, up to $^{238}_{92}U$, so the question remains as to how they are formed. It is thought that the production of heavier elements comes primarily from neutron capture reactions on ^{56}Fe (the heaviest, stable nucleus formed by heavy nucleus fusion reactions). For example, in a large neutron flux, the following reactions could occur,

$$^{56}Fe + n \rightarrow ^{57}Fe + \gamma \qquad (3.41)$$

$$^{57}Fe + n \rightarrow ^{58}Fe + \gamma \qquad (3.42)$$

$$^{58}Fe + n \rightarrow ^{59}Fe + \gamma \qquad (3.43)$$

Note that ^{56,57,58}Fe are all nuclei which are stable against radioactive decay, however, ^{59}Fe is β-unstable. Thus, whether ^{59}Fe takes on another neutron depends on (a) the density of the neutron flux and (b) the decay half-life of $^{59}_{26}$Fe, (45 days). If the neutron flux is too small, the ^{59}Fe will decay to the stable nucleus ^{59}Co which will subsequently capture a neutron to form the unstable system $^{60}_{27}$Co. However, if neutron density is high enough, $^{59}_{26}$Fe will capture another neutron before it decays to form $^{60}_{26}$Fe. This sequence of neutron captures will continue until an isotope is reached whose decay half-life is short enough for it to β-decay before another neutron can be captured. The process where the neutron flux is small and there is sufficient time for β-decay to occur before a further neutron is captured is known as the slow or s-process, while the second, faster process, where many neutrons can be captured before β-decay takes place is known as the rapid or r-process. The neutrons for the s-process are thought to come mainly from (α, n) reactions such as $^{13}C(\alpha, n)^{16}O$ and $^{22}Ne(\alpha, n)^{25}Mg$ which occur during the helium burning phases of second and third generation stars or in Red giants. For typical Red-giant internal temperatures of $1 \to 2 \times 10^8 K$ and estimated neutron densities (N) of $\sim 10^{14} m^{-3}$, the reaction rate per target atom (see later) is given by

$$r \simeq < \sigma v > N \qquad (3.44)$$

$T = 2 \times 10^8 K$ corresponds to a neutron velocity of around $2 \times 10^6 m s^{-1} (E_n \sim 20 keV)$. For typical neutron capture cross-sections (see later) around this energy of 100 mb, this corresponds to a rate of around 2×10^{-9} per atom per second, or about one capture every 30 years. Thus clearly, for the r-process to occur, much denser neutron fluxes must be present. In order to generate the vast, dense neutron flux required for the r-process, the flux density needs to be increased by around ten orders of magnitude!! One suggestion is that such a density may occur during a violent supernova. Close to the magic neutron shell at N=50, 82, 126, the γ-decay half-lives are so short compared even to the r-process neutron capture reaction times. This gives rise to the predicted linear increase in proton number for these magic neutron numbers (see Figure 3.9). These are known as waiting points in the -process. These nuclei subsequently γ-decay to form stable elements and this accounts for the observed increase in abundance for stable nuclei with A=80, 130 and 192 (corresponding to the magic neutron numbers at N=50, 82 and 126).

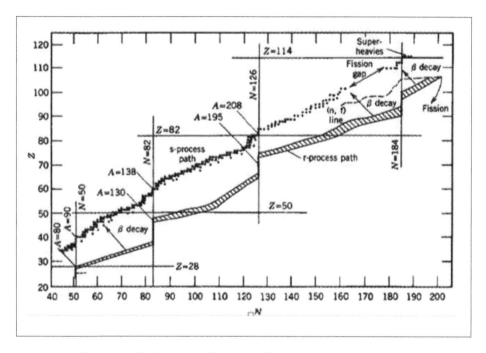

Figure 3.9: Predicted paths of the s and r-processes.

Solved Problems

1. Calculte the Q-value of the reaction $^7Li \rightarrow ^4He +^3H$. Given mass of the nuclei in units of a.m.u. as $M(^7Li)$=7.01822, $M(^4He)$=4.00387, $M(^3H)$=3.01700. Comment on the stability of the parent nucleus against the decay via the given reaction.

   Solution $Q = M(^7Li) - \{M(^4He) + M(^3H)\}$={7.01822-(4.00387+3.01700)} a.m.u.=(7.01822-7.02087)a.m.u = - 0.00265 a.m.u. = -2.468475 MeV
   Since Q < 0, the parent nucleus ^7Li is stable against the decay.

2. Estimate the energy released in the fusion of four proton (4H) into ^{4}He. Given:$M(^4He)$=4.002604 a.m.u., M(H)=1.007826 a.m.u. Use the data to estimate energy content of 1kg of hydrogen. Comment on the result in the light of energy production in the sun.

   Solution E=4M(H)-M(^{4}He) = (4× 1.007826 - 4.002604) a.m.u. = 0.0287 a.m.u. = 26.73405 MeV = 4.277 ×10^{-12}J.
   Number of protons in 1kg of hydrogen is approximately 6×10^{26}.

Therefore, the energy content of 1kg of hydrogen will be about

$$\frac{4.277 \times 10^{-12} J/4}{6 \times 10^{26}} \simeq 6.42 \times 10^{14} J$$

The corresponding energy radiation capacity will be about $2.04 J kg^{-1} s^{-11}$ for about 10^7 years.

Comment: Nuclear reactions leading to the fusion of four hydrogen nuclei to produce one helium nucleus could be sufficient for the energy production in the sun.

3. Calculate the Q-value of the reactions (a) $^4He +^4 He \to^8 Be$. Given: $\Delta^2(^4He)=+2.425$MeV, $\Delta(^8Be) = +4.942$MeV, 1u $=931.5$ MeV$/c^2 = 1.66 \times 10^{-24} gm$). [Hint: For the reaction $A + B \to C + D$, Q=Δ(A) + Δ(B)-Δ(C)-Δ(D)]

   Solution Q=$\Delta(^8Be) - 2 \times \Delta(^4He)$=(+4.942MeV)-2×(+2.425MeV) = +0.092MeV As Q<0, the product ^{8}Be is formed in the resonant state at resonance energy of 92keV. It is to be noted that for the reverse reaction Q>0, hence ^{8}Be is unstable aginst α decay.

Exercise

1. Show that the Q-value of the reaction $^3He +^4 He \to^7 Be$ is -92keV. Given: $k = 1.38 \times 10^{-23}$J/K, $\hbar = 1.05 \times 10^{-34}$Js, 1u=$1.66 \times 10^{-27}$kg $= 931.5$MeV$/c^2$, M(^{4}He)=4.002603u, M(^{3}He) = 3.016029u,(M(^{7}Be) = 7.016928u.

2. Discuss the difference between the simple CN and hot or β-limited CNO cycles. What reactions set the limit of rate of conversion of 4 protons to an α-particle in each case? What nuclei are formed in the hot CNO cycle that is not present in the simple CN cycle.

3. Using the fact that, three-fourths of the solar mass of $1.99 \times 10^{30} kg$ consists of protons, calculate the length of time that fusion energy can be generated at the present rate of $1.4 kW/m^2$ at a distance of $1.50 \times 10^{11} m$ in converting four protons to a ^{4}He nucleus.

---

[1]The sun having mass $2 \times 10^{30} kg$ with total energy output of $4 \times 10^{26} J$ per sec has a radiation capacity of $2 \times 10^{-4} J kg^{-1} s^{-1}$

[2]Note: $\Delta = $ M - A$m_u = $ mass excess, where M=atomic mass, A=mass number, m_u is the atomic mass unit=931.5meV$/c^2$. It is to be noted that for ^{12}C, mass excess is zero, i.e. $\Delta(^{12}C) = 0$

4. Calculate the energy released in each one of the nuclear reactions in all three PP- chains.

5. Using the Saha equation, estimate the ratio of number of ^{8}Be nuclei to ^{4}He nuclei for temperatures of $10^7, 10^8$ and $10^9 K$. (Assume an α-particle density of $3 \times 10^5 g/cm^3$).

6. How can nuclei with A>56 be formed in stars?

7. Show that the maximum amount of energy released in the form of electromagnetic radiation from converting four protons to a ^{4}He is given by the binding energy of ^{4}He less twice the sum of the neutron-proton mass difference and the mass of positrons. Ignore any rest mass the neutrino may have.

References

Quoted in the text:

- A.E.Champagne and M.Wiescher, Annual Review of Nuclear and Particle Science 42 (1992) p39-76.

- Käppler, Thielemann and Wiescher, Ann. Rev. Nucl. Part. Sci. 48 (1998) p175-251.

- H. Schatz et al. Physics Reports 294 (1998) p167-264).
  Further readings:

- Burbidge, E. M., Burbidge, G. R., Fowler, W. A., and Hoyle, F.: 1957, Rev. Mod. Phys. 29, 547.

- Fowler, W. A., Caughlan, G. R., and Zimmerman, B. A.: 1967, Ann. Rev. Astron. Astroph. 5, 525.

- Fowler, W. A., Caughlan, G. R., and Zimmerman, B. A. : 1975, Ann. Rev. Astron. Astroph. 13,69.

- Reeves, H.: 1964, Stellar Evolution and Nucleosynthesis, Gordon and Breach Science Pubs., Inc., New York.

Chapter 4

Opportunities in Nuclear Astrophysics

4.1 Introduction

It is an exciting time for nuclear astrophysics. There are opportunities to reach a new level of understanding of nucleosynthesis in the big bang, of the evolution of stars, and of explosive events such as supernovae, novae, x-ray-bursts, x-ray pulsars and neutron star mergers. An exquisitely detailed record of these events will flow from new astronomical observatories. Nuclear physics is so inextricably involved in astronomical phenomena, however, that only with a much better knowledge of nuclei and nuclear reactions can we obtain a deep understanding of these phenomena. Fortunately, promising new facilities exist or are on the horizon. New radioactive beam facilities will produce elements previously made only in stars and elucidate those of their properties important for cosmic phenomena. Stable beam accelerators with exceptional intensity and cleanliness will study nuclear reactions at energies close to those found in stellar environments. Powerful neutron sources will delineate processes that create the heavy elements. More powerful detectors of neutrinos from the sun and supernovae will provide information on neutrino properties, and on the role of neutrinos in explosive processes. With wise investment of our resources, great strides in our knowledge of the cosmos should be possible in the near future. We can anticipate finding the sources of energy density in the universe, based on nucleosynthesis in the big bang, measurements of the cosmic background radiation, and measurements of the acceleration of the cosmic expansion, using Type Ia supernovae as standard candles. We will probably have a solution of the solar neutrino problem and a picture of the nature of neutrinos: the number of neutrino types and their masses. We can expect to have assembled the requisite combination of computational power and nuclear physics knowledge to model supernova explosions. We should know where the heavy elements are formed and understand the processes that make them. We should have understood the complex mixing and mass loss dynamics, which accompanies late stellar evolution. We should have an improved understanding of the structure of neutron stars and of how their properties are modified by nuclear burning on their surfaces in binary systems. With luck, we will have constructed a detector for supernova neutrinos and have observed a (rare) supernova explosion. We can anticipate further challenges. Given the explosive growth of terrestrial and satellite based observatories, it seems certain that new phenomena will be observed and that new knowledge of nuclear physics will be crucial to understanding them.

4.2 Thermonuclear rates and reaction networks

This section highlights the nucleosynthesis processes in stellar evolution and stellar explosions, with an emphasis on the role of nuclei far from stability with a short introduction of the physics in astrophysical plasmas which governs composition changes. Few basic equations for thermonuclear reaction rates, nuclear reaction networks and burning processes with required nuclear physics input is discussed for cross sections of nuclear reactions, photo disintegrations, electron and positron captures, neutrino captures, inelastic neutrino scattering, and for beta-decay half-lives. Sub-barrier fusion reaction is also discussed in this section.

4.2.1 Thermonuclear Reaction Rates

The nuclear cross section for a reaction between projectile x and target y is defined by

$$
\begin{aligned}
\sigma &= \frac{number\ of\ reactions\ per\ sec\ per\ unit\ target\ y}{flux\ of\ incoming\ projectiles\ x} \\
&= \frac{n_r/n_y}{n_x v}
\end{aligned}
\tag{4.1}
$$

The second equality holds for the case that the relative velocity between targets with the number density n_y and projectiles with number density n_x is constant and has the value v. Then n_r, the number of reactions per cm^3

per sec, can be expressed as $n_r = \sigma v n_y n_x$. More generally, when targets and projectiles follow specific distributions, n_r is given by

$$(n_r)_{y,x} = \int \sigma(\vec{v_y} - \vec{v_x})|(\vec{v_y} - \vec{v_x})|d^3n_y d^3n_x. \tag{4.2}$$

The evaluation of this integral depends on the type of particles and distributions which are involved.

Maxwell-Boltzmann Distributions

For nuclei y and x in an astrophysical plasma, obeying a Maxwell-Boltzmann distribution,

$$d^3n_y = n_y(\frac{m_y}{2\pi kT})^{3/2}\exp(-\frac{m_y v_y^2}{2kT})d^3v_y \tag{4.3}$$

Eq. (4.2) is then simplified to

$$(n_r)_{y,x} = < \sigma v >_{y,x} n_y n_x \tag{4.4}$$

The thermonuclear reaction rates have the form

$$\begin{aligned} < y, x > &= < \sigma v >_{y,x} \\ &= (8\mu\pi)^{1/2}(kT)^{-3/2}\int_0^\infty E\sigma(E)\exp(-\frac{E}{kT})dE. \end{aligned} \tag{4.5}$$

Where μ denotes the reduced mass of the target-projectile system and the integral extends over the projectile energy range. In astrophysical plasmas with high densities and/or low temperatures, effects of electron screening become highly important. This means that the reacting nuclei, due to the background of electrons and nuclei, feel a different Coulomb repulsion than in the case of bare nuclei in a vacuum. Under most conditions (with non-vanishing temperatures) the generalized reaction rate integral can be separated into the traditional expression without screening and a screening factor

$$< y, x >^* = f_{scr}(Z_y, Z_x, \rho, T, Y_i) < y, x > \tag{4.6}$$

This screening factor is dependent on the charge of the involved particles, the density, temperature, and the composition of the plasma. Here Y_i denotes the abundance of nucleus i defined by $Y_i = n_i/(\rho N_A)$, where n_i is the number density of nuclei per unit volume and N_A, Avogadro number. At high densities and low temperatures screening factors can enhance reactions by many orders of magnitude and lead to psycho nuclear ignition. In the extreme case of very low temperatures, where reactions are only possible via ground state oscillations of the nuclei in a Coulomb lattice, Eq. (4.6) breaks down, because it was derived under the assumption of a Boltzmann distribution.

Planck Distributions in Photo disintegrations

When in Eq. (4.2) particle x is a photon, the relative velocity is always c and quantities in the integral are independent of $d^3 n_y$, which simplifies n_r to $(n_r)_y = \lambda_{y,\gamma} n_y$ and $\lambda_{y,\gamma}$ results from an integration of the photo disintegration cross section over a Planck distribution for photons of temperature T

$$d^3 n_\gamma = \frac{1}{\pi^2 (c\hbar)^3} \frac{E_\gamma^2}{\exp(E_\gamma/kT) - 1} dE_\gamma \tag{4.7}$$

$$(n_r)_y = \lambda_{j,\gamma}(T) n_y = \frac{\int d^3 n_y}{\pi^2 (c\hbar)^3} \int_0^\infty \frac{c\sigma(E_\gamma) E_\gamma^2}{\exp(E_\gamma/kT) - 1} dE_\gamma \tag{4.8}$$

There exist a number of recent attempts to evaluate experimental photo disintegration cross sections and determine photo disintegration rates. Due to detailed balance[1] these rates can also be expressed by the capture cross sections for the inverse reaction $l + m \to y + \gamma$ via

$$\lambda_{y,\gamma}(T) = \left(\frac{G_l G_m}{G_y}\right) \left(\frac{A_l A_m}{A_y}\right)^{3/2} \left(\frac{m_u kT}{2\pi\hbar^2}\right)^{3/2} < l, m > \exp\left(-\frac{Q_{lm}}{kT}\right) \tag{4.9}$$

This expression depends on the reaction Q-value Q_{lm}, the inverse reaction rate <l,m>, the partition functions $G(T) = \sum_i (2y_i + 1) \times \exp(-E_i/kT)$ and the mass numbers A of the participating nuclei in a thermal bath of temperature T.

Fermi Distributions in Weak Interactions.

A procedure similar to Eqs. (4.7) & (4.8) is used for electron captures on nuclei $e^- + (Z, A) \to (Z - 1, A) + \nu_e$. Because the electron is about 2000 times less massive than a nucleon, the velocity of the nucleus j is negligible in the center of mass system in comparison to the electron velocity ($|\vec{v}_y - \vec{v}_e| \simeq |\vec{v}_e|$) and the integral does not depend on $d^3 n_y$. The electron capture cross section has to be integrated over a Boltzmann, partially degenerate, or Fermi distribution of electrons, dependent on the astrophysical conditions. The electron capture rates are a function of T and $n_e = Y_e \rho N_A$, the electron number density. In a neutral, completely ionized plasma, the electron abundance is

---

[1]The principle of detailed balance is formulated for kinetic systems which are decomposed into elementary processes like collisions, or steps, or elementary reactions: At equilibrium, each elementary process should be equilibrated by its reverse process.

equal to the total proton abundance in nuclei

$$Y_e = \sum_i Z_i Y_i \qquad (4.10)$$

$$(n_r)_y = \lambda_{y,e}(T, \rho Y_e) n_y \qquad (4.11)$$

Theoretical investigations extended from simpler approaches to full shell model calculations for the involved Gamow-Teller and Fermi transitions in weak interaction reactions. The same authors generalized this treatment for the capture of positrons, which are in a chemical equilibrium with photons, electrons, and nuclei. Recent experimental results from charge-exchange reactions like $(d,^2 He)$ show a good agreement with theory. At high densities $(\rho > 10^{12} gcm^{-3})$ the size of the neutrino scattering cross section on nuclei and electrons ensures that enough scattering events occur to thermalize a neutrino distribution. Then also the inverse process to electron capture (neutrino capture, i.e. charged-current neutrino scattering) can occur and the neutrino capture rate can be calculated by integrating over the neutrino distribution. It is also possible that a thermal equilibrium among neutrinos was established at a different location than at the point where the reaction occurs. In such a case the neutrino distribution can be characterized by chemical potential at temperature which is not necessarily equal to the local temperature.

Decays

Finally, for normal decays, like β or α decays with half-life $\tau_{1/2}$, we obtain an equation similar to (4.7-4.8) or (4.11) with a decay constant

$$\lambda_y = \frac{ln2}{\tau_{1/2}} \qquad (4.12)$$

$$(n_r)_y = \lambda_y n_y \qquad (4.13)$$

Beta-decay half-lives $\tau_{1/2}$ for unstable nuclei can either be obtained from experiments or can be predicted theoretically in a way similar to quasi particle RPA calculations.

4.2.2 Nuclear Reaction Networks

The time derivative of the number densities of each of the species in an astrophysical plasma (at constant density) is governed by the different expressions for n_r, the number of reactions per cm^3 per s, as discussed above for the different reaction mechanisms which can change nuclear abundances

$$(\frac{\partial n_i}{\partial t})_{\rho=const} = \sum_y N_y^i (n_r)_y + \sum_{y,x} N_{y,x}^i (n_r)_{y,x} + \sum_{y,x,l} N_{y,x,l}^i (n_r)_{y,x,l}. \quad (4.14)$$

The reactions listed on the right hand side of the equation belong to the three categories of reactions: i) decays, photodisintegrations, electron and positron captures and neutrino induced reactions $((n_r)_y = \lambda_y n_y)$, ii) two particle reactions $(r_{y,x} =< y, x > n_y n_x)$, and (iii) three-particle reactions $((n_r)_{y,x,l} =< y, x, l > n_y n_x n_l)$ like the triple-alpha process, which can be interpreted as successive captures with an intermediate unstable target. The individual N^i's are given by:

$$N_y^i \quad = \quad N_i, N_{y,x}^i = N_i/\Pi_{m=1}^{n_m} |N_{ym}|!, N_{y,x,l}^i = N_i/\Pi_{m=1}^{n_m} |N_{ym}|!.$$

The N_i's can be positive or negative numbers and specify how many particles of species i are created or destroyed in a reaction. The denominators, including factorials, run over the nm different species destroyed in the reaction and avoid double counting of the number of reactions when identical particles react with each other (for example in the $^{12}C +^{12} C$ or the triple-alpha reaction). In order to exclude changes in the number densities $\dot{n}_i$, which are only due to expansion or contraction of the gas, the nuclear abundances $Y_i = n_i/(\rho N_A)$ were introduced. For a nucleus with atomic weight $A_i, A_i Y_i$ represents the mass fraction of this nucleus, therefore $\sum A_i Y_i = 1$. In terms of nuclear abundances Y_i, a reaction network is described by the following set of differential equations

$$\dot{Y}_i \quad = \quad \sum_y N_y^i \lambda_y Y_y + \sum_{y,x} N_{y,x}^i \rho N_A < y, x > Y_y Y_x$$

$$+ \quad \sum_{y,x,l} N_{y,x,l}^i \rho^2 N_A^2 < y, x, l > Y_y Y_x Y_l \quad (4.15)$$

Eq. (4.15) derives directly from (4.14) when the definition for the Y_i is introduced. This set of differential equations is solved with a fully implicit treatment. Then the stiff set of differential equations can be rewritten as difference equations of the form $\Delta Y_i/\Delta t = f_i(Y_y(t + \Delta t))$, where $Y_i(t + \Delta t) = Y_i(t) + \Delta Y_i$. In this treatment, all quantities on the right hand side are evaluated at time $t + \Delta t$. This results in a set of non-linear equations for the new abundances $Y_i(t + \Delta t)$, which can be solved using a multi-dimensional Newton-Raphson iteration procedure[2]. The total energy generation per gram, due to nuclear reactions in a time step Δt which

---

[2]J. Comp. Appl. Math. (JCAM) 109, 321 (1999)

changed the abundances by ΔY_i, is expressed in terms of the mass excess $M_{ex,i}c^2$ of the participating nuclei

$$\Delta\epsilon \;=\; -\sum_i N_A M_{ex,i} c^2 \Delta Y_i \tag{4.16}$$

$$\dot{\epsilon} \;=\; -\sum_i \dot{Y}_i N_A M_{ex,i} c^2 \tag{4.17}$$

As noted above, the important ingredients to nucleosynthesis calculations are decay half-lives, electron and positron capture rates, photo disintegrations, neutrino induced reaction rates, and strong interaction cross sections. For a number of explosive burning environments the understanding of nuclear physics far from stability and the knowledge of nuclear masses is a key ingredient. In recent years new Hartree-Fock- Bogoliubov or relativistic mean field approaches are addressing this question. Presently the FRDM model[3] still seems to provide the best reproduction of known masses.

4.2.3 Burning Processes in Stellar Environments

Nucleosynthesis calculations can in general be classified into two categories: I. nucleosynthesis during hydrostatic burning stages of stellar evolution on long timescales and II. nucleosynthesis in explosive events (with different initial fuel compositions, specific to the event). In the following we want to discuss shortly reactions of importance for both conditions and the major burning products.

Hydrostatic Burning Stages in Stellar Evolution

The main hydrostatic burning stages and most important reactions are:

- **H-burning:** there are two alternative reaction sequences, the different pp chains which convert ^{1}H into ^{4}He, initiated by $^1H(p,e^+\nu_e)^2$H, and the CNO cycle which converts ^{1}H into ^{4}He by a sequence of (p,γ) and (p,α) reactions on C, N, and O isotopes and subsequent beta-decays. The CNO isotopes are all transformed into ^{14}N, due to the fact that the reaction $^{14}N(p,\gamma)^{15}$O is the slowest reaction in the cycle.

- **He-burning:** the main reactions involved are the triple-α reaction ^{4}He$(2\alpha,\gamma)^{12}$C and $^{12}C(\alpha,\gamma)^{16}$O.

---

[3]P. Möller et al.: At. Data Nucl. Data Tables 59, 185 (1995)

- **C-burning:** $^{12}C(^{12}C, \alpha)^{20}$Ne and ^{12}C$(^{12}C, p)^{23}$Na. Most of the ^{23}Na nuclei will react with the free protons via $^{23}Na(p, \alpha)^{20}$Ne.

- **Ne-Burning:** $^{20}Ne(\gamma, \alpha)^{16}$O, $^{20}Ne(\alpha, \gamma)^{24}$Mg and $^{24}Mg(\alpha, \gamma)^{28}$Si. It is important that photo disintegrations start to play a role when $30kT \simeq Q$ (as a rule of thumb), with Q being the Q-value of a capture reaction. For those conditions sufficient photons with energies >Q exist in the high energy tail of the Planck distribution. As $^{16}O(\alpha, \gamma)^{20}$Ne has an exceptionally small Q-value of the order 4 MeV, this relation holds true for $T > 1.5 \times 10^9 K$, which is the temperature for (hydrostatic) Ne-burning.

- **O-burning:** The main reactions are $^{16}O(^{16}O, \alpha)^{28}$Si, $^{16}O(^{16}O, p)^{31}$P, and $^{16}O(^{16}O, n)^{31}S(\beta^+)^{31}$P. Similar to carbon burning, most of the ^{31}P is destroyed by a (p, α) reaction to ^{28}Si.

- **Si-burning:** Si-burning is initiated like Ne-burning by photo disintegration reactions which then provide the particles for capture reactions. It ends in an equilibrium abundance distribution around Fe (thermodynamic equilibrium).

As this includes all kinds of Q-values (on the average 8-10 MeV for capture reactions along the valley of stability), this translates to temperatures in excess of $3 \times 10^9 K$, being larger than the temperatures for the onset of Neburning. In such an equilibrium (also denoted nuclear statistical equilibrium, NSE) the abundance of each nucleus is only governed by the temperature T, density ρ, its nuclear binding energy B_i and partition function

$$G_i = \sum_y (2J_y^i + 1) \exp(-E_y^i/kT)$$

$$Y_i = (\rho N_A)^{A_i-1} \frac{G_i}{2^{A_i}} A_i^{3/2} \left(\frac{2\pi\hbar^2}{m_u kT} \right)^{\frac{3}{2}(A_i-1)} e^{\frac{B_i}{kT}} Y_p^{Z_i} Y_n^{N_i} \quad (4.18)$$

while fulfilling mass conservation $\sum_i A_i Y_i = 1$ and charge conservation $\sum_i Z_i Y_i = Y_e$ (the total number of protons equals the net number of electrons, which is usually changed only by weak interactions on longer timescales). This equation is derived from the relation between chemical potentials (for Maxwell-Boltzmann distributions) in a thermal equilibrium ($\mu_i = Z_i \mu_p + N_i \mu_n$), where the subscripts n and p stand for neutrons and protons. Intermediate quasi-equilibrium stages (QSE), where clusters of neighboring nuclei are in relative equilibrium via neutron and proton reactions, but different clusters have total abundances which are offset from their NSE values, are important during the onset of Si-burning before a full NSE is reached and during the freeze-out from high temperatures, which will be discussed in Chapter 5.

s-process

The slow neutron capture process leads to the build-up of heavy elements during core and shell He-burning, where through a series of neutron captures and beta-decays, starting on existing heavy nuclei around Fe, nuclei up to Pb and Bi can be synthesized. The neutrons are provided by a side branch of He-burning, $^{14}N(\alpha, \gamma)^{18}F(\beta^+)^{18}O(\alpha, \gamma)^{22}Ne(\alpha, n)^{25}$Mg. An alternative stronger neutron source in He-shell flashes is the reaction $^{13}C(\alpha, n)^{16}$O, which requires admixture of hydrogen and the production of ^{13}C via proton capture on ^{12}C and a subsequent beta-decay. Extensive overviews exist over the major and minor reaction sequences in all burning stages in massive stars. For less massive stars which burn at higher densities, i.e. experience higher electron Fermi energies, electron captures are already important in O-burning and lead to a smaller Ye or larger neutron excess $\eta = \sum_i (N_i - Z_i)Y_i = 1 - 2Y_e$. Most reactions in hydrostatic burning stages proceed through stable nuclei. This is simply explained by the long timescales involved. For a $25M_\odot$ star, which is relatively massive and therefore experiences quite short burning phases, this still amounts to: H-burning 7×10^6 y, He-burning 5×10^5y, C-burning 600 y, Ne-burning 1 y, O-burning 180 d, Si-burning 1 d. Because all these burning stages are long compared to beta-decay half-lives, with a few exceptions of long-lived unstable nuclei, nuclei can decay back to stability before undergoing the next reaction. Examples of such exceptions are the s-process branching with a competition between neutron captures and beta-decays of similar timescales.

Nuclear Burning in Explosive Events

Many of the hydrostatic burning processes discussed above can occur also under explosive conditions at much higher temperatures and on shorter timescales (see Figures 5.1 and 5.2 in the next chapter). The major reactions remain still the same in many cases, but often the beta-decay half-lives of unstable products are longer than the timescales of the explosive processes under investigation. This requires in general the additional knowledge of nuclear cross sections for unstable nuclei. Extensive calculations of explosive carbon, neon, oxygen, and silicon burning, appropriate for supernova explosions, have already been performed in the late 60s and early 70s with the accuracies possible in those days and detailed discussions about the expected abundance patterns. Besides minor additions of ^{22}Ne after He-burning (or nuclei which originate from it in later burning stages), the fuels for explosive nucleosynthesis consist mainly of alpha-particle nuclei like ^{12}C, ^{16}O, ^{20}Ne, ^{24}Mg, or ^{28}Si. Because the timescale of explosive processing is very short (a fraction of a second to several seconds), in most cases only few beta-decays can occur during explosive nucleosynthesis events, resulting in heavier nuclei, again with $N \simeq Z$. However, even for a fuel with a total N/Z $\simeq 1$ (or $Y_e \simeq 0.5$) a spread of nuclei around a line of N=Z is involved and many reaction rates for unstable nuclei have to be known. Depending on the temperature, explosive burning produces intermediate to heavy nuclei. The individual burning processes will be discussed chapter 5. Two processes differ from the above scenario for initial fuel compositions with extreme overall N/Z-ratios, where either a large supply of neutrons or protons is available, the r-process and the rp-process, denoting rapid neutron or proton capture (the latter also termed explosive hydrogen burning). The proton supply in the rp-process results from the accretion of unburned H and He in novae and X-ray bursts. Electron captures in supernova explosions of both types (Ia and II) can reduce Ye drastically, i.e. enhance the overall N/Z ratio. Neutrino-induced reactions can act in addition in type II supernovae. The astrophysical site which provides the neutron-rich conditions for the r-process is still debated, involving type II supernovae or neutron star mergers. In the r- or rp-process nuclei close to the neutron and proton drip lines can be produced and beta-decay timescales can be short in comparison to the process timescales.

4.3 Experimental Techniques in Nuclear Astrophysics

The techniques used for experimental study of nuclear observables of astrophysical interest are very varied depending on the stellar environment under investigation which could be quiescent or explosive burning. For illustration experimental techniques typically used in nuclear astrophysics studies are discussed in following few sections.

4.3.1 Choice of Target

The measurement of reaction cross sections at very low energies with intense stable-isotope beams (of few mA) and water-cooled thick solid targets requires the monitoring of the stability and of the stoichiometry of the target as a function of the beam current. This is typically performed by measuring from time to time the cross section at a reference energy (target stability) and by nuclear reaction analysis before and after the target has been employed (target stoichiometry). The appropriate corrections have to done in the experimental results[4].The study of reactions involved in explosive astrophysical processes need the development of radioactive beams. Reactions involving H and He are the most important ones in explosive burning, since the lifetime of the interacting nuclei are very small enough to allow for production of targets, inverse kinematics methods using radioactive beams require the use of H or He rich targets. The choice of the target must be based on the physics goal and the other experimental conditions like the beam energy and intensity, detection arrangements, etc. Polyethylene foils $(CH_2)_n$ are easy to handle and have been one of the most popular and successful targets for investigation of hydrogen burning reactions. Foils having thicknesses in the range $40 \mu g cm^{-2}$ to several $mg cm^{-2}$ have been used with beam intensities as high as $10^9 s^{-1}$[5] without significant degradation, though care has to be ensured to distribute the beam power by say, rotating the target. Solid targets containing helium are produced by implantation technique developed at Louvain-la-Neuve with helium thicknesses up to $10^{18} atoms/cm^2$, sufficient for measurement of elastic scattering and some reactions cross sections with radioactive ion beams[6]. Gas targets are an obvious alternative to foils. Gas cells with thin windows are easy to handle, but the windows produce similar challenges as with foil targets, degrading the beam energy and inducing background reactions. Windowless gas targets eliminate the problems associated with windows. However, large number of pumping stations are required to decrease the pressure to the desired range of 10^{-7}mbar, necessary for conducting on-line experiments at accelerators. Thus the gas targets are large, costly and have limited target thickness.

4.3.2 Choice of detectors

Studies of capture reactions of interest in quiescent burning are mainly limited by the cosmic background of the γ detectors. One can build a lead wall around the detectors but the interaction of the cosmic rays with the material will produce γ-rays and neutrons that will also affect the measurements. Another possibility to partially reduce the activation problem

[4]C. Iliadis: Nuclear Physics of Stars (Wiley-VCH Verlag GmbH, 2007) 254, 258, 260, 261

[5]W. Galster et al.: Phys. Rev. C 44, 2776 (1991) 263

[6]F. Vanderbist et al.: Nucl. Instr. and Meth. in Phys. Res. B 197, 165 (2002) 263

Table 4.1: Relation between astrophysical sites and nuclear processes.

Stellar Site	Nuclear Process
Big Bang (Primordial Nucleosynthesis)	**Reaction between light elements** p, d, He, Li
Main sequence stars (Eg. the Sun)	**Hydrogen burning** Proton-proton chains, CNO cycle, Ne-Na cycle, Mg-Al chain
Red giant stars Asymptotic giant branch (AGB) stars	**Helium burning** tripleα process,$^{12}C(\alpha, \gamma)^{16}O$, Other (α, γ) and (α, n) reactions
Super giant stars Wolf-Rayet stars, Pre-supernova	**Advance burning stages** Reactions of C, O, N, Ne, Si,....
Nova Supernova x-ray bursts	**Explosive burning** Hot CNO cycle Rapid proton capture (rp- process)
AGB stars Supernova type -II Neutron stars	**Nucleosynthesis beyond iron** Slow neutron capture (s-process) Rapid neutron capture (r-process) Photodisintegration and proton capture (p-process)

is an active shielding using, for example, plastic scintillators operated in anti-coincidence with the γ detectors. One of the best solution is going underground[7] but that is not always possible. The pioneers of underground laboratories for nuclear astrophysics reaction measurements is the LUNA laboratory, situated under the Gran Sasso mountain in Italy. Its unique character is a suppression of the cosmic rays equivalent to 4km of water. Two linear accelerators that are installed at LUNA (50 and 400 keV) have allowed to measure cross sections of the order of 0.01 pb. An example of these measurements at the limits of the technical possibilities is the study of the ^{14}N$(p, \gamma)^{15}$O reaction[8]. In the United States, a project for the construction of an underground laboratory has been launched by a collaboration of several universities and laboratories from the US and Europe. Experimental studies with radioactive beams of low intensities ($\sim 10^4 - 10^8 s^{-1}$ on target) suggest the use of very efficient detection systems. Arrays of γ, neutron and charged-particle detectors have been constructed at many facilities in order to maximize the detection efficiency[9]. The development of large-area silicon strip detectors covering a large solid angle has played a crucial role. These detectors may be segmented in one or two dimensions (doubled-sided detectors) to any practical level of pixilation. The shape of the strips can also be modified to satisfy experimental requirements. For example, strips are curved in a circular pattern in many annular detector designs to allow better reaction angle resolution in a strip. This approach was used in the Louvain–Edinburgh detector array (LEDA), one of the pioneering charged-particle arrays used with radioactive ion beams for nuclear astrophysics[10]. LEDA is composed of independent 16-strip sectors. It's typical electronic resolution is of the order of 10 keV, the energy resolution for 5.5 MeV α particles is about 20 keV, while the time resolution is of about 1 ns. Figure 4.1 shows a schematic drawing of one of the LEDA sectors and a typical experimental setup using two LEDA arrays. A broader range of detector thicknesses has recently become available, and detectors between 50 μm and 1 mm are common. This broad range of thicknesses allows Z identification of a broad range of charged particles through$\Delta E - E$ techniques. These detectors needs new associated electronic modules and new data acquisition systems capable to work with a large number of signal channels and

---

[7]G. Fiorentini, R.W. Kavanagh, C. Rolfs: Z. Phys. A 350, 289 (1995) 264

[8]A. Formicola et al.: Phys. Lett. B 591, 61 (2004) 260, 264, 271, 272

[9]J.C. Blackmon, C. Angulo, A.C. Shotter: Nucl. Phys. A 777, 531 (2006) 255, 256, 264

[10]T. Davinson et al.: Nucl. Instr. and Meth. in Phys. Res. A 454, 350 (2000) 264, 265, 269

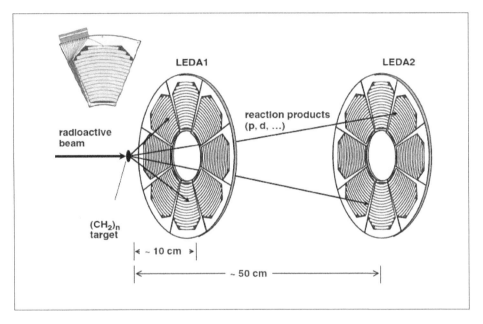

Figure 4.1: Typical experimental setup for the measurement of a transfer reaction by means of the LEDA detector. The insert on the top left is a schematic drawing of one LEDA sector (See also Plate 30 in the Color Plate Section); Ref. Nucl. Instr. and Meth. in Phys. Res. A 454, 350 (2000) 264

a small dead time. Also because of the very low-beam intensities it is not possible to use the conventional methods to measure the accumulated beam dose with sufficient precision. In the measurement of cross sections or of resonance strengths, the absolute normalization is always one of the most important tasks. One of the techniques successfully employed is to evaporate a very thin gold layer on the target (when using a plastic foil as a target) and normalize to the measurement of the Rutherford cross section. Arrays of γ-ray detectors have played an important role in several new approaches using both stable-isotope and radioactive ion beams. For instance, the high-total efficiency of the Gamma sphere array [11] allowed $\gamma - \gamma$ coincidence measurements. They accurately determined excitation energies of levels in proton-rich nuclei that are of astrophysical importance. Some of the problems with direct measurements, mainly the low efficiency of γ-detectors, the radioactivity of the target material, and the background sources can be solved by performing measurements in inverse kinematics and detecting the recoiling reaction products in recoil separators.

4.3.3 Recoil separators

Recoil separators are devices which separate the nuclear reaction products (re- coils) leaving the target from the primary beam and focus the former onto a detector system[12]. The experimental challenge is to maintain a high transmission of the heavy reaction recoils while maximizing the rejection of the primary beam. This is difficult because of the small mass and momentum difference between the projectiles and recoils. It is also difficult because all the projectiles enter the separator, their intensity being typically $10^{10} - 10^{15}$ times larger than of the recoils. To obtain the maximum separation, the primary beam is blocked at an early stage of the separator. Some recoil separators have the additional property of dispersing the reaction products at the focal plane according to their mass-to-charge ratio. The scattered beam rejection is enhanced by filtering out particles on the basis of both velocity and ratio between both mass and ionic charge, which necessitates the use of either velocity filters combined with magnetic dipoles or a combination of electric and magnetic dipole elements. A recoil separator suitable for (p,γ) and (α, γ) studies in astrophysics should have the following specifications[13]:

- High-beam rejection over a broad mass range of beams.

---

[11]I.Y. Lee: Nucl. Phys. A 520, 641c (1990) 26567

[12]C.N. Davids: Nucl. Instr. and Meth. in Phys. Res. B 204, 124 (2003) 266

[13]C.N. Davids: Nucl. Instr. and Meth. in Phys. Res. B 204, 124 (2003) 266

- High-transport efficiency for a relative small solid angle (typically less than 5 msr): due to the inverse kinematics, the maximum angles of the ejectiles should peak near $0°$

- Relative low-mass resolution ($\delta M/M \leq 0.5\%$).

- Incorporation of careful beam handling upstream of the separator (e.g. clean recoil beam with small dispersion and no beam halo).

- Target chamber capable of accommodating a variety of detector arrays and gas targets (both jet and extended targets).

- Capability of running with different ion optical modes for reactions with different kinematics.

Several laboratories have developed recoil separators that are designed to collect heavy reaction products and disperse them by their mass-to-ionic-charge ratio. Such separators for astrophysical studies are, for example, DRAGON at ISAC (TRIUMF), ERNA at Bochum, the Dares bury Recoil Separator at the HRIBF (Oak Ridge, and the FMA at ANL (Argonne). A modern separator dedicated to nuclear astrophysics studies using stable-isotope beams is under construction at the University of Notre Dame[14].

4.3.4   Ground State Properties

Knowledge of β-decay half lives and masses of nuclei are important ingredients for the understanding of explosive processes. Half lives for β decay can be long compared to the time scale for nuclear reactions, and thus the decays of nuclei near the proton drip line can govern energy generation and nucleosynthesis in the explosion. As the rates of nuclear reactions depend exponentially on the reaction Q value, mass measurements are a crucial first step towards determining these rates. The development of highly selective spectrometers, traps, detectors, and other instrumentation at laboratories around the world has allowed nearly all isotopes of interest for explosive hydrogen burning to be produced and identified. Half lives and masses, which can be measured with relatively few atoms, are often the first quantities determined experimentally. there are few particle-stable neutron-deficient isotopes with $Z < 53$ whose half lives have not been measured with reasonable accuracy. Most important for the rp-process are the last unobserved even-even nuclei below ^{100}Sn, like ^{74}Sr, ^{78}Zr (N = Z - 2) and ^{96}Cd (N = Z)

---

[14]J. Görres: Proceedings of Science, (NIC-IX) 027. 266

which could have reasonably long half lives. Although the general progress in studying half lives in this region of the chart of nuclei is indeed impressive, there still exist many isotopes heavier than nickel that lack accurate mass measurements or other experimental information. The techniques developed for capturing and cooling nuclei in traps and storage rings should allow the situation with nuclear masses to be much improved in the near future. However, the additional structure information necessary to extract reaction rates will require substantially more effort.

4.3.5   Resonances Properties

Some reaction rates at the temperatures of explosive burning are totally or partially dominated by the contribution of resonances. It is therefore important to study the properties of the resonant states using, e.g. elastic and/or inelastic scattering, transfer reactions populating the mirror states, and fusion evaporation reactions as discussed below:

Elastic and Inelastic Scattering

Since the middle of the 20th century[15], elastic scattering method being is a popular choice to study resonant states. As a natural extension, the elastic scattering technique in inverse kinematics, used normally to investigate reactions involving radioactive species, makes use of the sensitivity of the protons (or α particles) to the presence of a resonant state in the compound nucleus. The principle of the scattering method is demonstrated in Figure 4.2. The technique is based on the fact that the energy loss of heavy ions in a target is significantly larger than the energy loss of protons which is normally negligible for typical target thicknesses of less than 1–2 mg/cm^2. The resulting recoil proton spectrum can be compared to a "snapshot" of a certain energy range in the level scheme of the compound nucleus. The method can also be applied to α scattering, although being less sensitive. Another advantage is that the laboratory energy of the recoil particles (protons or α particles) are rather high and given by:

$$E_{lab} = E_{CM} \frac{4A_x}{A_x + A_X} \cos^2 \phi_{lab} \qquad (4.19)$$

where A_x, A_X are the mass number of the projectile and the target nuclei, respectively, and ϕ_{Lab} is the angle of recoil. On the other hand, the cross section for elastic scattering is proportional to the square of two contributions,

---

[15]R.A. Laubestein et al.: Phys. Rev. 84, 12 (1951) 267

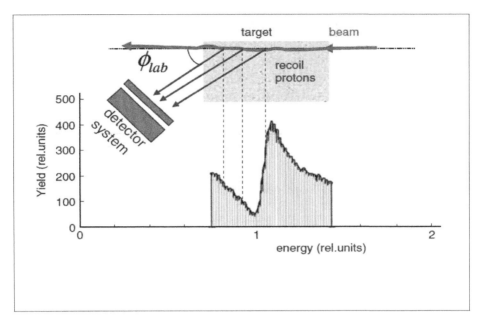

Figure 4.2: Principles of the elastic scattering method in inverse kinematics. The spec- trum is a typical interference pattern for a $l = 0$ resonance. Ref:Th. Delbar et al.: Nucl. Phys. A 542, 263 (1992) 267

the Coulomb amplitude and the nuclear amplitude[16]:

$$\frac{d\sigma}{d\Omega} = |f_{Nuc} + f_{Coul}|^2 \qquad (4.20)$$

The nuclear amplitude (f_{Nuc}) depends on the collision matrix $C_{lI}^{J\pi}$ that can be written as a function of the phase shift

$$C_{lI}^{J\pi} = \exp(2i\delta_{lI}^{J\pi}) \qquad (4.21)$$

where J and π are respectively the spin and parity of the resonant state, I the channel spin and l the angular momentum. When a resonant state is "scanned" with the appropriate combination projectile–target, the recoil proton spectra show an interference pattern that is indicative of the resonance energy, angular momentum and width (see Figure 4.2). All experimental effects, mainly the beam energy resolution inside the target and the angular resolution of the detectors, must be properly taken into account to precisely extract the resonant properties. Fitting procedures, such as the R-matrix

---

[16]A.M. Lane, R.G. Thomas: Rev. Mod. Phys. 30, 257 (1958) 268

method, are typically used to evaluate these quantities and to obtain the resonance properties. Elastic scattering measurements with radioactive beams have been widely applied to determine resonant properties that are important for nuclear astrophysics and nuclear structure. One of the main advantages is the large cross section for elastic scattering that allows measurements with radioactive beam intensities as low as $10^4 s^{-1}$. Another advantage is that the energy of the recoil protons is usually sufficiently high to be detected readily in standard silicon detectors or in more sophisticated strip detector arrays. The β radioactivity of the beam species may induce background in the particle detectors at low energies. This background can limit the lowest energies measurable, but this time-uncorrelated background can be distinguished from the events of interest by time-of-flight techniques. An example of an elastic scattering measurement of astrophysical interest is the study $^{21}Na + p$ at the ISAC facility at TRIUMF. An intense $(5^7 s^{-1})^{21}$Na beam at laboratory energies below 1.5 MeV/nucleon bombarded 50 - 250 $\mu g cm^{-2}$ thick $(CH_2)_n$ targets. The recoil protons were detected using the TUDA strip detector array.

4.3.6 Transfer Reactions

Much of our understanding of nuclear spectroscopy has been shaped by the study of transfer reactions in the last decades. These have proved to be a particularly powerful tool for characterizing energy levels of importance for astrophysical reactions. One advantage of these measurements is that states covering a broad region of excitation energy are populated. The main disadvantage from the experimental point of view is that this method needs high-resolution particle detection in order to resolve the states of interest. In addition, difficult targets and problematic kinematics make studies of resonant states via proton transfer in inverse kinematics particularly challenging. On the theoretical side, many of these reactions can be described by the distorted-wave born approximation (DWBA) which predicts that the shape of the angular distribution of the differential cross section is distinctive of the transferred angular momentum, and the magnitude of the cross section reflects the single-particle character of the state[17]. However, the results are usually model-dependent. Proton-transfer reactions, for example $(^3He, d)$, are the best surrogate for proton-induced reactions like (p, γ). The (d, p) neutron-transfer reaction has also been used to indirectly obtain information on single-particle resonances. The properties of neutron single-particle

---

[17]G.R. Satchler: Direct Nuclear Reactions (Clarendon Press, Oxford, 1983) 270

states are studied by the (d, p) reaction on the mirror nucleus, and the properties of proton resonances are determined under the assumption of mirror symmetry. This technique was first applied with a radioactive ion beam to ^{56}Ni at Argonne National Laboratory by Rehm et al.[18] and it has been in use to study the $^{18}F(p, \alpha)^{15}O$ at Louvain-la-Neuve and at the HRIBF. More exotic reactions, such as $(^{3}He,^{8}Li)$, $(^{4}He,^{8}He)$, $(^{7}Li,^{8}He)$ and $(^{12}C,^{6}He)$, have also been used to study states of nuclei further away from stability by means of stable-isotope beams and targets. The mechanisms of such reactions are complex, and the cross sections are typically small. However, in some cases high-intensity stable-isotope beams can be used to achieve reasonable reaction yields. States of unnatural parity are sometimes populated with comparable yields to natural parity states, allowing states to be studied that are weakly populated in direct reactions. For example, states of unnatural parity in ^{26}Si, that dominate the rate of the $^{25}Al(p, \gamma)^{26}Si$ reaction and are thus important for understanding the production of ^{26}Al in novae, were studied using the $^{29}Si(^{3}He,^{6}He)^{26}Si$ reaction at Wright Nuclear Structure Laboratory (WNSL) at Yale University[19]. Nuclei heavier than nickel in the rp-process are too far away from stable nuclei to be produced by transfer reactions involving stable-isotope beams and targets. Radioactive ion beams are thus required to access these nuclei by transfer reactions. Fragmentation facilities like that at the Michigan State University (MSU) produce nuclei near the rp-process path as beams with sufficient intensity to allow transfer reaction studies. Nucleon knock-out re- actions like (p, d) have much more favorable cross sections than stripping reactions at the beam energies available from fragmentation facilities. A set of measurements proposed for MSU will use the (p, d) reaction induced by a radioactive ion beam to populate proton-unbound states of nuclei near the path of the rp-process. The emitted protons and residual heavy nuclei being detected in coincidence in order to construct excitation energy spectra.

4.4 Experimental Nuclear Astrophysics

Experimental nuclear astrophysics provides reaction or decay data for simulating nucleosynthesis or energy generation patterns in stable or explosive stellar scenarios. These data are mostly determined in accelerator based experiments in the energy range between 6 keV and 60 GeV, depending on the goal of the experiment. Experimental nuclear astrophysics therefore

---

[18]Phys. Rev. Lett. 80, 676 (1998) 270
[19]J.A. Caggiano et al.: Phys. Rev. C 65 055801 (2002) 270

uses an enormous range of challenging experimental techniques. While the field has been extremely successful over the last few decades in testing and confirming reaction and decay rate predictions for the astrophysics community, it also posed new questions and challenges for this field by determining experimentally reactions which had been neglected or ignored for general nucleosynsthesis considerations. Presently experimental nuclear astrophysics is facing three major challenges,

- the study of very low energy reactions at stability to interpret charged particle induced nucleosynthesis processes during stellar evolution

- the measurement of neutron induced processes near stability for the understanding of the s-process

- the study of reaction and decay processes near and far from stability for interpretation of stellar explosion scenarios and the associated nucleosynthesis patterns.

The first subject is mainly concerned with nuclear reactions with charged particles of relevance for stellar hydrogen, helium, and carbon burning. While many of these reactions have been studied over the last four decades, most of the measurements were focused on energies well above the critical energy range of the Gamow window (see Sect. 4.3) because of the rapidly decreasing cross sections towards low energies. The presently used reaction rates (tabulated for example in the NACRE compilation) are mostly based on extrapolation of the higher energy cross sections towards the Gamow range. This procedure can carry considerable uncertainty, in particular when unaccounted for near threshold states can contribute as resonances to the reaction rate. The measurement of neutron capture reactions is of relevance for the study of the s-process during stellar core He burning in massive stars ($M \geq 15 M_{\odot}$) via weak s-process and in AGB stars ($M \leq 4 M_{\odot}$) via strong s-process. Most of the measurement have been done with the activation technique using neutron beams with an energy distribution simulating a 25 keV Maxwell Boltzmann distribution. New detector technology allowed to also perform in-beam γ spectroscopy. These measurements were often handicapped by the need for large sample masses, which prohibited the study of neutron capture on rare isotopes. Alternative techniques were the use of electron beam induced pulsed neutron sources like ORELA with time of flight analysis. The neutrons are produced by bremsstrahlung from a tantalum radiator. Neutron spallation sources like n-ToF at CERN provide higher intensity white neutron beams which now allow in combination with time of flight analysis to measure the neutron capture cross sections on considerably smaller samples of rare isotopes than previously possible. In this volume we will concentrate on the third item, the important experimental questions and challenges presented by the goal to study reactions or decay processes far off stability using radioactive beam techniques.

4.4.1   Energy ranges for measurement of cross-sections

The nuclear reaction rate per particle pair at a given stellar temperature T is determined by folding the reaction cross section with the Maxwell-Boltzmann (MB) velocity distribution of the projectiles, as demonstrated in Eq. (4.5). Two cases have to be considered, reactions between charged particles and reactions with neutrons. The nuclear cross section for charged particles is strongly suppressed at low energies due to the Coulomb barrier. For particles having energies less than the height of the Coulomb barrier, the product of the penetration factor and the MB distribution function at a given temperature results in the so-called Gamow peak, in which most of the reactions will take place. Location and width of the Gamow peak depend on the charges of projectile and target, and on the temperature of the interacting plasma. When introducing the astrophysical S-factor

$$S(E) = \sigma(E)Ee^{2\pi\eta}$$

(with η being the Sommerfeld parameter, describing the s-wave barrier penetration), one can easily see the two contributions of the velocity distribution and the penetrability in the integral

$$< \sigma v >= \left(\frac{8}{\pi\mu}\right)^{1/2} \int_0^\infty S(E) \exp\left(-\frac{E}{kT} - \frac{b}{E^{1/2}}\right) dE \qquad (4.22)$$

where the quantity $b = 2\pi\eta E_{1/2} = (2\mu)^{1/2}\pi e^2 Z_y Z_x/\hbar$ arises from the barrier penetrability. Taking the first derivative of the integrand yields the location E_0 of the Gamow peak, and the effective width Δ of the energy window can be derived accordingly

$$E_0 = (bkT/2)^{2/3} = 1.22(Z_y^2 Z_x^2 A T_6^2)^{1/3} keV \qquad (4.23)$$

$$\Delta = \frac{16E_0 kT^{1/2}}{3} = 0.749(Z_y^2 Z_x^2 A T_6^5)^{1/6} keV \qquad (4.24)$$

where the charges Z_y, Z_x, the reduced mass A of the involved nuclei in units of m_u, and the temperature T_6 given in $10^6 K$, at the center. In the case of neutron-induced reactions the effective energy window has to be derived in a slightly different way. For s-wave neutrons ($l = 0$) the energy window is simply given by the location and width of the peak of the MB distribution function. For higher partial waves the penetrability of the centrifugal barrier shifts the effective energy E_0 to higher energies. For neutrons with energies less than the height of the centrifugal barrier this was approximated as

$$E_0 \simeq 0.172 T_9 (l + \frac{1}{2}) MeV \qquad (4.25)$$

$$\Delta \simeq 0.194 T_9 (l + \frac{1}{2})^{1/2} MeV \qquad (4.26)$$

The energy E_0 will always be comparatively small in comparison to the neutron separation energy.

4.4.2 Radioactive Beams

Radioactive beam experiments are necessary for the measurement of reactions and decay processes of radioactive nuclei which can take place at the high temperatures, typical in explosive stellar events. At these conditions the Gamow window represents a much higher energy than for hydrostatic burning stages in stellar evolution. The reaction rates, which for charged particle interactions typically increase exponentially with temperature, become much larger than the decay rates and the reaction path runs far away from the line of stability. Nucleosynthesis simulations of explosive scenarios up to now are mostly based on theoretical predictions for capture and decay rates far off stability. These theoretical input parameters need to be confirmed or complemented by reliable reaction and decay rate measurements. Within the last decade reaction measurements with radioactive beams have been successfully performed, simulating nuclear processes in the Big Bang and for explosive hydrogen and helium burning conditions typical for the thermonuclear runaway[20] in accreting binary star systems or for explosive burning in supernova shock fronts. For neutron induced processes such as the r-process, the particular neutron capture reaction rates are less important since the r-process path is essentially determined by an $(n, \gamma) - (\gamma, n)$ chemical equilibrium which depends on the nuclear masses. Therefore an increasing number of studies has focused on the global properties of neutron rich nuclei such as masses, half-lives and neutron decay probabilities. All these are important ingredients for r-process nucleosynthesis models. Neutron capture rates itself will be difficult if not impossible to measure but

---

[20]The Astrophysical Journal, 568:779–790, 2002

recent activities focus on the study of neutron transfer measurements on radioactive neutron rich isotopes instead to probe the level density and single particle structure of neutron unbound states in nuclei towards the r-process path.

ISOL Facilities.

The development of clean radioactive beams presents an enormous technical challenge. For most of the low energy experiments mentioned below, the so-called Isotope Separation On-Line (ISOL) technique was applied where the radioactive beam particles were produced by nuclear reactions or spallation processes with high energy primary stable beams in a production target. The fast release of the produced radioisotopes is important for achieving high beam intensities but requires optimization of the chemical and physical characteristics of the isotope/target combination. The released isotopes need to be separated by electric and magnetic mass/charge separation systems and post-accelerated up to the required beam energies. ISOL facilities (like e.g. Louvain-la-Neuve, CERN-Isolde, the Oak-Ridge Holifield Facility) have been used over the years as an excellent tool for the production of radioactive isotopes for decay measurements along the r-process and the rp-process path. Capture measurements with radioactive beams in inverse kinematics are among the major goals for radioactive beam facilities. The first pioneering measurements on university based radioactive beam facilities operated with limited beam intensities (on average $\leq 10^7 ions/s$ with some beams reaching exceptional intensities of up to 10^9 ions/s). These limitations are due to the cross sections of production reactions, the release time for the radioactive species in the production target and finally the chemistry conditions in the ion source. The experiments were also handicapped by insufficient detector systems with relatively low detection efficiency and limited background reduction capabilities. Nevertheless, these first attempts proved that radioactive beam experiments for nuclear astrophysics can be successfully performed and can provide important data for the nuclear astrophysics community. With the above constraints in mind, only reactions with high cross sections have been successfully measured such as $^{13}N(p,\gamma)$, $^{8}Li(\alpha,n)$, $^{18}F(p,\alpha)$, and $^{18}Ne(\alpha,p)$. The experimental conditions in terms of detector development have improved continuously. The increasing use of large Si-array strip detector systems for low energy (p,α) and (α,p) experiments such as LEDA was a breakthrough in terms of efficiency and solid angle and helped to compensate for the limited beam intensities available. The use of these detectors allowed to investigate resonance states for capture reactions through elastic resonance scattering if the resonance capture cross section was too low for a direct measurement. These elastic scattering studies were in particular successful for $^{17}F(p,p)$ where they helped to uniquely identify the energy of a missing resonance state. Experimental conditions have been further improved by the design and utilization of recoil-mass-separators. These guarantee a $\leq 10^{-12}$ rejection of the primary beam and further background reduction by particle identification methods while maintaining a high detection efficiency. This was clearly demonstrated in the first successful experiment of low energy resonances in $^{21}Na(p,\gamma)$ with the DRAGON separator[21]. Recoil mass separators are also now being commissioned or constructed at other low energy radioactive ion beam facilities such as Holifield Radioactive Ion Beam Facility (HRIBF) and Louvain la Neuve. These, coupled to a powerful γ-detection array will dominate the future measurements with low energy radioactive beams. A new generation of ISOL based radioactive beam facilities is emerging with theISAC facility at TRIUMF, Rex-Isolde at CERN, and Spiral at GANIL. They will hopefully provide higher beam intensities of up to $10^{10} - 10^{12} atoms/s$, depending on the chemical and physical characteristics of the isotope. Optimization of beam intensity mainly requires new and innovative developments in target technology to reduce the ionization[22] and extraction times of the radioactive isotopes. The first capture reaction measurement of $^{21}Na(p,\gamma)$ has been successfully completed at ISAC using a surface ionization source. This experiment concentrated on the measurement of two low energy resonances at 205 keV and 821 keV. In particular the lower energy resonance controls the reaction rate at nova temperatures and is therefore critical for the understanding of the production of the long-lived radionuclide ^{22}Na in novae. Future measurements of $^{19}Ne(p,\gamma)$ and $^{18}Ne(\alpha,p)$ are planned using an ECR source. These reactions are closely associated with the break-out from the hot CNO cycles. Developing the experimental technique and in measuring the higher energy resonances at Louvain la Neuve were successful. The study of the resonances within the Gamow range of X-ray burst conditions (see Chapter 8) requires higher beam intensities and novel detection and background reduction schemes to improve over the previously determined upper limits. Similar experiments are planned at Rex-Isolde like the study of resonances in $^{35}Ar(p,\gamma)^{36}K$, a reaction that controls the reaction flow in

---

[21]Z. Y. Bao, H. Beer, F. Käppeler et al.: At. Data Nucl. Data Tables 76, 70 (2000)

[22]Ionization is the process by which an atom or a molecule acquires a negative or positive charge by gaining or losing electrons to form ions, often in conjunction with other chemical changes. There are different techniques of ionization, such as MALDI: matrix assisted laser desorption ionization.

high temperature novae burning and may limit the production of ^{40}Ca in novae. A simple but efficient technique is the so-called in-beam production of radioactive isotopes which has been the backbone of the radioactive beam programs at Notre Dame and Argonne. The radioactive particles are produced by heavy ion nuclear reactions and separated from the primary beam in flight by the magnetic field structure of superconducting solenoids. The energy of the secondary beam is mainly determined by the kinematics of the production process. The beam intensity is typically limited depending on the cross section of the production process.

4.4.3   In-Flight Separators.

In-flight separators present an alternative approach for the production of neutron–rich or neutron–deficient radioactive isotopes between the region of stability and the neutron or proton drip–line. They are mainly based on separating (in flight) high energy radioactive heavy ion reaction or fission products through mass separator systems from the primary beam component (see also the lecture by Morrissey and Sherrill). Depending on magnetic, electric, and absorption conditions in the separator a so-called cocktail beam is produced at the focal plane which consists of particle groups within a certain A,Z range. The various cocktail components are identified by a subsequent energy, energy-loss analysis. Due to the large momentum transfer in the initial production process these beams have a fairly high energy and are typically used for mass or half–life measurements. Increasingly also transfer reactions (on stable and radioactive beams) are performed to study the evolution of nuclear structure far from stability. Such nuclear structure results are extremely important since they often provide important complementary information for determining the reaction rates along the r– or rp–process path. Important impact for the description of the r-process but also the rp-process has been provided by the measurements of β-decay or β-delayed particle decay (emission) processes along the process path. In particular these measurements presented a major challenge for studying β-decays and β-delayed neutron emission for r-process nuclei which were typically located far from stability. In most of these cases the isotopes of interest were separated by ISOL or in-flight separator techniques and implanted into periodically moving tapes to be detected off-line. Pulsed beam techniques from in-flight separator facilities typically implanted the separated short-lived particles into a stack of Si-detectors after particle identification to monitor the decay on-line. These measurement may provide a more sensitive approach for identifying radioactive isotopes far off stability and for reducing the background which often limited the accessibility to very neutron rich isotopes at ISOL beam facilities. Those measurements were often handicapped by background from long lived daughter activities of isobaric impurities in the separated particle groups. Only in a few cases direct measurements on r-process nuclei have been successful like the study of ^{130}Cd in 1986 at ISOLDE/CERN, which later could be significantly improved with the use of better isobar separation. In particular the development and use of laser ion sources opened new opportunities for removing isobaric impurities from the beam, reducing this kind of background enormously. Fragment separator beams have been used in particular for life time and decay studies of radioactive nuclei near the rp-process or r-process path. The high energy radioactive isotopes are directly implanted into stacks of Si detectors or into moving tape systems to measure the accumulated activity off-line. In the recent past in-flight separators have been successfully used to map both masses and life times near the proton drip line. In particular the systematic study of lifetimes in the mass A=35 to 65 along the drip line has provided important information for rp-process simulations and put the previously used theoretical estimates on firm experimental ground. Detailed studies at NSCL/MSU, ORNL and GANIL have focused on the N=Z nuclei above A=64 to investigate the waiting point nuclei $^{64}Ge, ^{68}Se$, and ^{72}Kr up to ^{80}Zr. These measurements did focus mainly on the lifetime and decay properties of these isotopes but provided also important information about the masses through β-decay endpoint measurements. The use of traps can and will considerably improve the accuracy of such studies in the near future. While most of the measurements near the r-process path have been performed at ISOL based facilities, the upgrade of in-flight separator facilities like the coupled cyclotron facility at NSCL/MSU opened new opportunities to provide access to r-process nuclei, first measurements have been successfully completed in the lower mass range of the r-process, preferably in the study of the decay properties of neutron rich Ni isotopes. These measurements concentrated both on the study of the β decay as well as the β-delayed neutron decay properties of these isotopes. In view of the recent successes of γ-ray astronomy and the anticipated results of the recently launched new INTEGRAL observatory, life time measurements of long-lived isotopes which contribute to the observed galactic radioactivity are of particular importance. These kind of measurements have been tedious with classical radiochemical methods and have been haunted by considerable experimental uncertainties. The on-line production and counting of long-lived radioactive isotopes at in-flight separator facilities provides an alternative experimental tool to reduce the uncertainties considerably. This laboratory information does provide an important test for model predictions about the actual production rates for long-lived radioactive nuclei in the associated nucleosynthesis event. High energy radioactive beams from in-flight separators are obviously not usable for low energy capture measurements. But they have been used for Coulomb dissociation studies to obtain the capture cross section by applying the detailed balance theorem[23]. In particular measurements of $^8B(\gamma, p)^7Be$ have been performed at Notre Dame, Riken, GSI Darmstadt, and the NSCL at Michigan State University to determine the $^7Be(p, \gamma)^8B$ reaction rate. This reaction is critical for the production of high energy neutrinos in our sun and the rate is important for determining the neutrino oscillation parameters. Coulomb dissociation was successfully applied for cases were the cross section was dominated by a single resonance like in $^{13}N(p, \gamma)^{14}O$ or the direct capture like in $^7Be(p, \gamma)^8B$, but the interpretation of the Coulomb dissociation spectrum becomes difficult as soon as several resonance and/or reaction components contribute. It also requires a strong known ground state transition branch of the resonance decay for converting correctly the cross section by detailed balance.

4.5   Cross Section Predictions and Reaction Rates

Explosive nuclear burning in astrophysical environments produces unstable nuclei, which again can be targets for subsequent reactions. In addition, it involves a very large number of stable nuclei, which are not and cannot be fully explored by experiments. Thus, it is necessary to be able to predict reaction cross sections and thermonuclear rates with the aid of theoretical models. Explosive burning in supernovae involves in general intermediate mass and heavy nuclei. Due to a large nucleon number they have intrinsically a high density of excited states. A high level density in the compound nucleus at the appropriate excitation energy allows to make use of the statistical model approach for compound nuclear reactions which averages over resonances. Here, we want to present recent results obtained within this approach and outline in a clear way, where in the nuclear chart and for which environment temperatures its application is valid. It is often colloquially termed that the statistical model is only applicable for intermediate and heavy nuclei. However, the only necessary condition for its application is a large number of resonances at the appropriate bombarding energies, so that the cross section can be described by an average over resonances. This can in specific cases be valid for light nuclei and on the other hand not be

---

[23]E. Caurier, K. Langanke, G. Martinez-Pinedo, F. Nowacki: Nucl. Phys. A 653, 439 (1999)

valid for intermediate mass nuclei near magic numbers. In astrophysical applications usually different aspects are emphasized than in pure nuclear physics investigations. Many of the latter in this long and well established field were focused on specific reactions, where all or most "ingredients", like optical potentials for particle and alpha transmission coefficients, level densities, resonance energies and widths of giant resonances to be implemented in predicting E1 and M1 gamma-transitions, were deduced from experiments. This of course, as long as the statistical model prerequisites are met, will produce highly accurate cross sections. For the majority of nuclei in astrophysical applications such information is not available. The real challenge is thus not the well established statistical model, but rather to provide all these necessary ingredients in as reliable a way as possible, also for nuclei where none of such informations are available. In addition, these approaches should be on a similar level as e.g. mass models, where the investigation of hundreds or thousands of nuclei is possible with manageable computational effort. The statistical model approach has long been employed in calculations of thermonuclear reaction rates for astrophysical purposes, which in the beginning only made use of ground state properties. Later, the importance of excited states of the target was pointed out . The compilations by permitted large scale applications in all subfields of nuclear astrophysics, when experimental information is unavailable. Existing global optical potentials, mass models to predict Q-values, deformations etc., but also the ingredients to describe giant resonance properties have been quite successful in the past. Besides necessary improvements in global alpha potentials, the major remaining uncertainty in all existing calculations stems from the prediction of nuclear level densities, calculations of which yield uncertainties even beyond a factor of 5-10 at the neutron separation energy. In nuclear reactions the transitions to low-lying states dominate due to the strong energy dependence. Because the deviations are usually not as high yet at lower excitation energies, the typical cross section uncertainties amounted to a smaller factor of 2–3. Some novel treatments for level density descriptions, where the level density parameter is energy dependent and shell effects vanish at high excitation energies, improves the level density accuracy. This is still a phenomenological approach, making use of a back-shifted Fermi-gas model rather than a combinatorial approach based on microscopic single-particle levels. But it was the first one leading to a reduction of the average cross section uncertainty to a factor of about 1.4, i.e. an average deviation of about 40% from experiments, when only employing global predictions for all input parameters and no specific experimental knowledge. Par and Pezer showed that combinatorial approach is equivalent to a back-shifted Fermi gas, provided the parameters are determined from consistent physics. Thus, it is not surprising that fully microscopic approaches give similar uncertainties.

4.5.1   Thermonuclear Rates from Statistical Model Calculations

A high level density in the compound nucleus permits to use averaged transmission coefficients T, which do not reflect a resonance behavior, but rather describe absorption via an imaginary part in the (optical) nucleon-nucleus potential leading to the well known expression

$$
\sigma_a^{\mu\nu}(b,o;E_{ab}) = \frac{\pi\hbar^2/(2\mu_{ab}E_{ab})}{(2J_a^\mu+1)(2J_b+1)}
$$
$$
\times \sum_{b,\pi}(2J+1)\frac{T_b^\nu(E,J,\pi,E_a^\mu,J_a^\mu,\pi_a^\mu)T_o^\nu(E,J,\pi,E_m^\nu,J_m^\nu,\pi_m^\nu)}{T_{tot}(E,J,\pi)}
\tag{4.27}
$$

for the reaction $a^\mu(b,o)m^\nu$ from the target state a^μ to the exited state m^ν of the final nucleus, with a center of mass energy E_{ab} and reduced mass μ_{ab}. J denotes the spin, E the corresponding excitation energy in the compound nucleus, and π the parity of excited states. When these properties are used without subscripts they describe the compound nucleus, subscripts refer to states of the participating nuclei in the reaction $a^\mu(b,o)m^\nu$ and superscripts indicate the specific excited states. Experiments measure $\sum_\nu \sigma_a^{o\nu}(b,o;E_{ab})$, summed over all excited states of the final nucleus, with the target in the ground state. Target states μ in an astrophysical plasma are thermally populated and the astrophysical cross section $\sigma_a^*(b,o)$ is given by

$$
\sigma_a^*(b,o;E_{ab}) = \frac{\sum_\mu(2J_a^\mu+1)\exp(-E_a^\mu/kT)\sum_\nu \sigma_a^{\mu\nu}(b,o;E_{ab})}{\sum_\mu(2J_a^\mu+1)\exp(-E_a^\mu/kT)}
\tag{4.28}
$$

The summation over ν replaces $T_o^\nu(E,J,\pi)$ in Eq. (4.27) by the total transmission coefficient

$$
T_0(E,J,\pi) = \sum_{\nu=0}^{\nu_m} T_0^\nu(E,J,\pi,E_m^\nu,J_m^\nu,\pi_m^\nu)
$$
$$
+ \int_{E_m^{\nu_m}}^{E-s_{m,o}} T_0(E,J,\pi,E_m,J_m,\pi_m)
$$
$$
\times \rho(E_m,J_m,\pi_m)dE_m
\tag{4.29}
$$

Here $S_{m,o}$ is the channel separation energy, and the summation over excited states above the highest experimentally known state ν_m is changed to an integration over the level density ρ. The summation over target states μ in Eq. (4.28) has to be generalized accordingly. It should also be noted at this point that the formulation given here assumes complete mixing of isospin. For reactions with $N = Z \pm 1$, i.e. isospin 0 or $\pm 1/2$ targets, deviations are observed but can be well reproduced in an appropriate approach. In addition to the ingredients required for Eq. (4.27), like the transmission coefficients for particles and photons, width fluctuation corrections $W(b, o, J, \pi)$ have to be employed. They define the correlation factors with which all partial channels for an incoming particle b and outgoing particle o, passing through the excited state (E, J, π), have to be multiplied. This takes into account that the decay of the state is not fully statistical, but some memory of the way of formation is retained and influences the available decay choices. The major effect is elastic scattering, the incoming particle can be immediately re-emitted before the nucleus equilibrates. Once the particle is absorbed and not re-emitted in the very first (pre-compound) step, the equilibration is very likely. This corresponds to enhancing the elastic channel by a factor W_b. In order to conserve the total cross section, the individual transmission coefficients in the outgoing channels have to be renormalized to T_b'. The total cross section is proportional to T_b and, when summing over the elastic channel $(W_b T_b')$ and all outgoing channels $(T_{tot}' - T_b')$, one obtains the condition $T_b = T_b'(W_b T_b'/T_{tot}') + T_b'(T_{tot}' - T_b')/T_{tot}'$. We can (almost) solve for T_b'

$$T_b' = \frac{T_b}{1 + T_b'(W_b - 1)/T_{tot}'} \tag{4.30}$$

This requires an iterative solution for T (starting in the first iteration with T_b and T_{tot}), which converges fast. The enhancement factor W_b has to be known in order to apply (4.30). A fit to results from Monte Carlo calculations gave

$$W_b = 1 + \frac{2}{1 + T_b^{1/2}} \tag{4.31}$$

Eqs. (4.30) and (4.31) redefine the transmission coefficients of Eq. (4.27) in such a manner that the total width is redistributed by enhancing the elastic channel and weak channels over the dominant one. Cross sections near threshold energies of new channel openings, where very different channel strengths exist, can only be described correctly when taking width fluctuation corrections into account. The width fluctuation corrections of are only an approximation to the correct treatment. The important ingredients of statistical model calculations as indicated in Eq. (4.27) through Eq. (4.29) are the particle and gamma-transmission coefficients T and the level density of excited states ρ. Therefore, the reliability of such calculations is determined by the accuracy with which these components can be evaluated (often for unstable nuclei). In the following we want to discuss the methods utilized to estimate these quantities and recent improvements.

Transmission Coefficients.

The transition from an excited state in the compound nucleus (E, J, π) to the state $(E_a^\mu, J_a^\mu, \pi_a^\mu)$ in nucleus i via the emission of a particle b is given by a summation over all quantum mechanically allowed partial waves

$$T_b^\mu(E, J, \pi, E_a^\mu, J_a^\mu, \pi_i^\mu) = \sum_{l=|J-s|}^{J+s} \sum_{s=|J_a^\mu - J_b|}^{J_a^\mu + J_b} T_{bls} E_{ab}^\mu \qquad (4.32)$$

Here the angular momentum $\vec{l}$ and the channel spin $\vec{s} = \vec{J_b} + \vec{J_a^\mu}$ couple to $\vec{J} = \vec{l} + \vec{s}$. The transition energy in channel b is $E_{ab}^\mu = E - S_b - E_a^\mu$. The individual particle transmission coefficients T_l are calculated by solving the Schrödinger equation with an optical potential for the particle-nucleus interaction. Most of the calculation of thermonuclear reaction rates employ optical square well potentials using the black nucleus approximation. However, Thielemann et al[24] employed the optical potential for neutrons and protons, based on microscopic infinite nuclear matter calculations for a given density, applied with a local density approximation[25]. Deformed nuclei can be treated in a very simplified way by using an effective spherical potential of equal volume, based on averaging the deformed potential over all possible angles between the incoming particle and the orientation of the deformed nucleus. In most earlier compilations alpha particles are treated by square well optical potentials. The gamma-transmission coefficients are included for the dominant gamma-transitions (E1 and M1) in the calculation of the total photon width. Whereas the smaller, and therefore less important, M1 transitions are usually treated with the simple single particle approach $T \propto E^3$. The E1 transitions are usually calculated on the basis of the Lorentzian representation of the Giant Dipole Resonance (GDR), where the E1 transmission coefficient for the transition emitting a photon of energy

[24]F.-K. Thielemann, M. Arnould, J. W.Truran: In: Advances in Nuclear Astrophysics, ed. E. Vangioni-Flam (Editions Fronti'ere, Gif sur Yvette 1987) p. 525

[25]For a general overview on different approaches see R.L. Varner, W.J. Thompson, T.L. McAbee, E.J. Ludwig, T.B. Clegg: Phys. Rep. 201, 57 (1991)

E_γ in a nucleus $^N_Z A$ is given by

$$T_{E1}(E_\gamma) = \frac{8}{3} \frac{NZ}{A} \frac{e^2}{c\hbar} \left(\frac{1+\chi}{mc^2}\right) \sum_{a=1}^{2} \frac{a}{3} \frac{\Gamma_{G,a}E_\gamma^4}{(E_\gamma^2 - E_{G,a}^2)^2 + \Gamma_{G,a}^2 E_\gamma^2} \qquad (4.33)$$

where $\chi(= 0.2)$ accounts for the neutron-proton exchange contribution and the summation over a includes two terms which correspond to the split of the GDR in statically deformed nuclei, with oscillations along (a=1) and perpendicular (a=2) to the axis of rotational symmetry. Many microscopic and macroscopic models have been devoted to the calculation of the GDR energies (E_G) and widths (Γ_G). But the hydrodynamic droplet model approach adopted by Thielemann for E_G gives an excellent fit to the GDR energies and can also predict the split of the resonance for deformed nuclei, when making use of the deformation, calculated within the droplet model. In that case, the two resonance energies are related to the mean value calculated by the relations $E_{G,1} + 2E_{G,2} = 3E_G$, $E_{G,2}/E_{G,1} = 0.911\eta + 0.089/$ Here η is the ratio of the diameter along the nuclear symmetry axis to the diameter perpendicular to it, and can be obtained from the experimentally known deformation or mass model predictions. An important aspect of the E1 strength is related to so-called pygmy resonances at low energies which can be investigated in the relativistic mean field approaches. The connection of energy and strength of the pygmy resonance as a function of isospin and neutron skin of nuclei is still not fully understood. However a full microscopic treatment with large scale QRPA (Quasi-particle Random Phase Approximation) calculations for the E1 strength performed by Goriely et al [26] has been implemented in statistical model calculations. The accuracy of the extrapolations towards the drip lines depends, similar to mass predictions, on the quality of the microscopic model applied.

Level Densities

In the determination of cross sections nuclear level density introduces the largest uncertainties. Hence for large scale astrophysical applications it is necessary to not only find reliable methods for level density predictions, but also computationally feasible ones. And non-interacting Fermi-gas model could be a good choice for this purpose, because most of the statistical model calculations use the back-shifted Fermi-gas description of Gilbert et al [27]. The more sophisticated Monte Carlo shell model calculations by Dean et al

[26] Nucl. Phys. A 706, 217 (2002)
[27] A. Gilbert, A.G.W. Cameron: Can. J. Phys. 43, 1446 (1965)

[28] as well as combinatorial approaches by Paar group[29] have justified the application of the Fermi-gas description in the range of neutron separation energies. Rauscher, Thielemann, and Kratz[30] applied an energy-dependent level density parameter d_l together with microscopic corrections from nuclear mass models leading to an improved fits in the mass range $20 \leq A \leq 245$. The back-shifted Fermi-gas description assumes an even distribution of odd and even parities.

$$\rho(U, J, \pi) = \frac{1}{2} F(U, J)\rho(U), \tag{4.34}$$

$$with\ \rho(U) = \frac{1}{\sqrt{2\pi}\sigma} \frac{\sqrt{\pi}}{12d_L^{1/4}} \frac{\exp(2\sqrt{d_L U})}{U^{5/4}},$$

$$F(U, J) = \frac{2J + 1}{2\sigma^2} \exp(\frac{-J(J + 1)}{2\sigma^2}) \tag{4.35}$$

$$\sigma^2 = \frac{\Theta_{rigid}}{\hbar^2} \sqrt{U/d_L}, \Theta_{rigid} = \frac{2}{5} m_\mu A R^2, U = E - \delta$$

Where the spin dependence function F is determined by the spin cut-off parameter σ. Thus, the level density depends on two parameters: the level density parameter d_L and the back shift δ, which determines the energy of the first excited state. Within this framework, the quality of level density predictions depends on the reliability of systematic estimates of d_L and δ. In their first compilation for a large number of nuclei Gilbert et al found that the back shift δ is well reproduced by experimental pairing corrections of Cameron et al[31]. Gilbert group was also first to identify an empirical correlation with experimental shell corrections $C_S(Z, N)$

$$\frac{d_L}{A} = c_0 + c_1 C_S(Z, N), \tag{4.36}$$

where $C_S(Z, N)$ becomes negative near shell closures. The back-shifted Fermi-gas approach diverges for $U = 0 (i.e. E = \delta$, if δ is a positive back-shift). In order to obtain the correct behavior at very low excitation energies, the Fermi-gas description can be combined with the constant temperature formula

$$\rho(U) \propto \frac{\exp(U/T)}{T} \tag{4.37}$$

[28]D.J. Dean, S.E. Koonin, K. Langanke, P.B. Radha, Y. Alhassid: Phys. Rev. Lett. 74, 2909 (1995)

[29]V. Paar, R. Pezer: Phys. Lett. B 411, 19 (1997)

[30]Phys. Rev. C 57, 2031 (1997)

[31]A.G.W. Cameron, R.M. Elkin: Can. J. Phys. 43, 1288 (1965)

The two formulations are matched by a tangential fit yielding the value of the transmission coefficient T. There have been a number of compilations for d_L and δ, or T, based on experimental level densities[32]. An improved approach has to consider the energy dependence of the shell effects, which are known to vanish at high excitation energies. Although, for astrophysical purposes only energies close to the particle separation thresholds have to be considered, an energy dependence can lead to a considerable improvement of the global fit. This is especially true for strongly bound nuclei close to magic numbers. An excitation-energy dependent description of Ignatyuk et al[33] for the level density parameter d_L followed

$$d_L(U, Z, N) = \tilde{d}_L(A)[1 + C_M(Z, N)\frac{f(U)}{U} \tag{4.38}$$

$$\tilde{d}_L(A) = \alpha A + \beta A^{2/3} \tag{4.39}$$

$$f(U) = 1 - \exp(-\gamma U) \tag{4.40}$$

The values of the free parameters α, β and γ are determined by fitting to experimental level density data available over the whole nuclear chart. The shape of the function f(U) permits the two extremes: (i) for small excitation energies the original form of Eq.(4.36) $d_L/A = \alpha + \alpha\gamma C_M(Z, N)$ is retained with $C_S(Z, N)$ being replaced by $C_M(Z, N)$, (ii) for high excitation energies d_L/A approaches the continuum value α, obtained for infinite nuclear matter. Neglecting β in both the cases is justified, since earlier attempts to find a global description of level densities used shell corrections C_S derived from comparison of liquid-drop masses with experimental values ($C_S \equiv M_{exp} - M_{LD}$) or the "empirical" shell corrections $C_S(Z, N)$. A problem connected with the use of liquid-drop masses arises from the fact that there are different liquid-drop model parameterizations available in the literature which produce quite different values for C_S[34]. However, in addition, the meaning of the correction parameter inserted into the level density formula given in Eq.(4.38) has to be reconsidered. The fact that nuclei approach a spherical shape at high excitation energies (temperatures) has to be included. Therefore, the correction parameter C should describe properties of a nucleus differing from the spherical macroscopic energy and contain those terms which are finite for low and vanishing at higher excitation energies. The latter requirement followed the form of Eq. (4.38). Therefore,

---

[32]see for example: T. von Egidy, H.H Schmidt, A.N. Behkami: Nucl. Phys. A 481, 189 (1988)

[33]A.V. Ignatyuk, K.K. Istekov, G.N. Smirenkin: Sov. J. Nucl. Phys. 29, 450 (1979)

[34]A. Mengoni, Y. Nakajima: J. Nucl. Sci. Techn. 31, 151 (1994)

the parameter $C_M(Z, N)$ should rather be identified with the so-called "microscopic" correction E_{mic} than with the shell correction. The mass of a nucleus with deformation can then be written in two ways

$$M(\epsilon) = E_{mic}(\epsilon) + E_{mac}(spherical) \tag{4.41}$$
$$M(\epsilon) = E_{mac}(\epsilon) + E_{s+p}(\epsilon), \tag{4.42}$$

with E_{s+p} being the shell-plus-pairing correction. This confusion over the term "microscopic correction", being sometimes used in an ambiguous way, has been addressed by Möller group.[35] The so called ambiguity follows from the inclusion of deformation-dependent effects into the macroscopic part of the mass formula. Another important ingredient is the pairing gap Δ, related to the back shift δ. Instead of assuming constant pairing[36] or an often applied fixed dependence on the mass number like $\pm 12/\sqrt{A}$, the pairing gap Δ can be determined from differences in the binding energies (or mass differences, respectively) of neighboring nuclei[37].

$$\Delta_n(Z, N) = 1/2[M(Z, N-1) + M(Z, N+1) - 2M(Z, N)], \tag{4.43}$$

where Δn is the neutron pairing gap and M(Z,N) the ground state mass excess of the nucleus (Z,N). Similarly, the proton pairing gap Δp can be calculated by

$$\Delta_p(Z, N) = 1/2[M(Z-1, N) + M(Z+1, N) - 2M(Z, N)], \tag{4.44}$$

This is still a phenomenological rather than a combinatorial approach based on microscopic single-particle levels. The combinatorial approach is equivalent to a back-shifted Fermi gas, provided the parameters are determined from consistent physics. Thus, it is not surprising that fully microscopic approaches give similar uncertainties. An important effect at low excitation energies can come from the parity distribution of levels, deviating from the assumption entering in Eq. (4.34).

Results:

Rauscher et al [38] utilized the microscopic corrections of the Finite Range Droplet Model (FRDM) mass formula invoking a folded Yukawa shell model

---

[35]P. Möller et al.: At. Data Nucl. Data Tables 59, 185 (1995)

[36]W. Reisdorf: Z. Phys. A 300, 227 (1981)

[37]R.-P. Wang, F.-K. Thielemann, D.H. Feng, C.-L. Wu: Phys. Lett. B 284, 196 (1992)

[38]T. Rauscher, F.-K. Thielemann: At. Data Nucl. Data Tables 75, 1 (2000); 79, 47 (2001)

with Lipkin-Nogami pairing in to determine the parameter $C(Z, N) = E_{mic}$. The back shift δ was calculated by setting

$$\delta(Z, N) = \frac{1}{2}\Delta n(Z, N) + \Delta p(Z, N) \qquad (4.45)$$

and using Eq. (4.44). The parameters α, β, and γ were obtained from a fit to experimental data for s-wave neutron resonance spacing of 272 nuclei at the neutron separation energy using data of Iljinov et al[39]. For a quantitative overall estimate of the agreement between calculations and experiments, one usually quotes the ratio

$$g \equiv \langle \frac{\rho_{calc}}{\rho_{exp}} \rangle \;\; = \;\; \exp[\frac{1}{n}\sum_{i=1}^{n}(ln\frac{\rho_{calc}^{i}}{\rho_{exp}^{i}})]^{1/2} \qquad (4.46)$$

with n being the number of nuclei for which level densities ρ are experimentally known. As best fit we obtain an averaged ratio $g = 1.48$ with the parameter values $\alpha = 0.1337, \beta = -0.06571, \gamma = 0.04884$. This corresponds to $d_L/A = \alpha = 0.134$ for infinite nuclear matter, which is approached for high excitation energies. The ratios of experimental to predicted level densities (i.e. theoretical to experimental level spacing D) for the nuclei considered are shown in Figure 4.3 which indicates an absolute deviation less than a factor of 2 for majority of nuclei. This is a satisfactory improvement over theoretical level densities used in previous astrophysical cross section calculations, where deviations of a factor 3-4, or even in excess of a factor of 10 were found. With these improvements, the uncertainty in the level density is now comparable to uncertainties in optical potentials and gamma transmission coefficients which enter the determinations of capture cross sections. The remaining uncertainty of extrapolations is the one due to the reliability of the nuclear structure model applied far from stability which provides the microscopic corrections and pairing gaps.

Applicability of the Statistical Model.

For a reliable application of statistical model, a sufficiently large number of levels in the compound nucleus is needed in the relevant energy range, which can act as doorway states to the formation of the compound nucleus. For narrow, isolated resonances, the cross sections (and also the reaction rates) can be represented by a sum over individual Breit-Wigner terms. The main question is whether the density of resonances (i.e. level density) is

---

[39] A.S. Iljinov et al.: Nucl. Phys. A 543, 517 (1992)

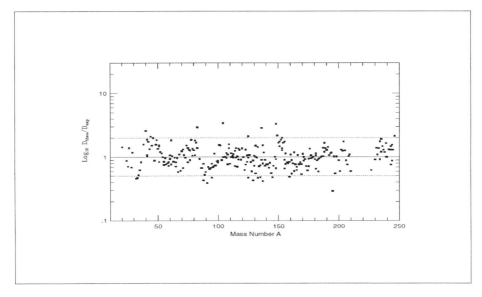

Figure 4.3: Ratio of predicted to experimental level densities at the neutron separation energy. The deviation is less than a factor of 2 (dotted lines) for the majority of the considered nuclei.

high enough so that the integral over the sum of Breit-Wigner resonances may be approximated by an integral over the statistical model expressions of (4.24), which assume that at any bombarding energy a resonance of any spin and parity is available. Numerical test calculations have been performed by Rauscher et al. (1997)[40] in order to find the average number of levels per energy window which is sufficient to allow this substitution in the specific case of folding over a MB distribution. To achieve 20% accuracy, about 10 levels in total are needed in the effective energy window in the worst case (non-overlapping, narrow resonances). This relates to a number of s-wave levels smaller than 3. Application of the statistical model for a level density which is not sufficiently large, results in general in an overestimation of the actual cross section, unless a strong s-wave resonance is located right in the energy window[41]. Thus one may assume a conservative limit of 10 contributing resonances in the effective energy window for charged and neutral particle-induced reactions. To obtain the necessary number of levels within the energy window of width Δ can require a sufficiently high excitation energy, as the level density increases with energy. This combines with

---

[40]Phys. Rev. C 57, 2031 (1997)
[41]See for details: L. van Wormer, et al. Astrophys. J. 432, 326 (1994)

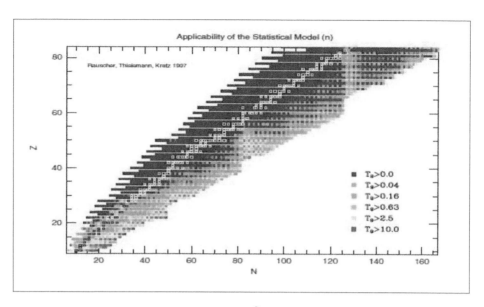

Figure 4.4: Stellar temperatures (in $10^9 K$) for which the statistical model can be used. Plotted is the compound nucleus of the neutron-induced reaction n+Target. Stable nuclei are marked in the same grey-scale but with a light grey frame.

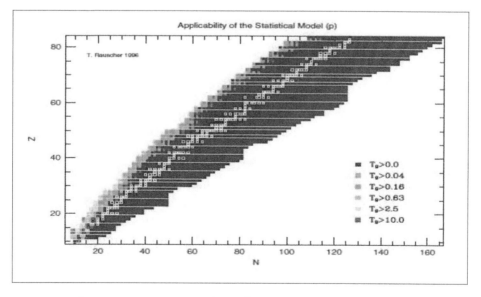

Figure 4.5: Stellar temperatures (in 10^9) for which the statistical model can be used. Plotted is the compound nucleus of the proton-induced reaction p+Target. Stable nuclei are marked as in the previous figure.

the thermal distribution of projectiles to a minimum temperature for the application of the statistical model. Those temperatures are plotted in a logarithmic grey scale in Figures 4.4 - 4.5. For neutron-induced reactions Figure 4.4 applies while Figure 4.5 describes proton-induced reactions. Plotted is always the minimum stellar temperature T_9 (in $10^9 K$) at the location of the compound nucleus in the nuclear chart. It should be noted that the derived temperatures will not change considerably, even when changing the required level number within a factor of about two, because of the exponential dependence of the level density on the excitation energy. This permits to read directly from the plot whether the statistical model cross section can be "trusted" for a specific astrophysical application at a specified temperature or whether single resonances or other processes (e.g. direct reactions) have also to be considered. These plots can give hints on when it is safe to use the statistical model approach and which nuclei have to be treated with special attention for specific temperatures. Thus, information on which nuclei might be of special interest for an experimental investigation may also be extracted. The general information can be taken that neutron induced reactions are problematic close to the neutron drip-line and proton induced reactions close to the proton drip-line. This is simply due to the very low excitation energy at which the compound nucleus is formed in such reactions. Alpha-induced reactions, not plotted here, have also very similar reaction Q-values across the nuclear chart and therefore it is in most cases safe to apply the statistical model.

4.5.2   Astrophysical S-factors of radiative capture reactions

In this section an outline to describe the astrophysical S-factors of radiative capture reactions with light atomic nuclei on the basis of the potential two-cluster model, in which interaction of the nucleon clusters can be described by local two-particle potential determined by fit to the scattering data and properties of bound states of these clusters is presented . The astrophysical S-factors of the radiative capture processes in the p^2H, p^7Li and p^{12}C systems can be analyzed on the basis of this approach. It can be seen that the approach allows one to describe quite reasonably experimental data available at low energies, when the phase shifts of cluster-cluster scattering are extracted from the data with minimal errors. The explanation of ways of the chemical element formation in the stars is one of the significant conclusions of the modern nuclear astrophysics. The nuclear doctrine of origin of elements describes the prevalence of different elements in the Universe on the basis of characteristics of these elements taking into account physical conditions in which they can be formed. In addition, the set of considering nuclear astrophysics processes allows to interpret, for example, the star luminosity on the different stages of their evolution and to describe in general outline the process of stellar evolution itself. Hereby, the nucleosynthesis questions are closely coupled, on the one part, with the questions of structure and evolution of the stars and the Universe and on the other part with the nuclear particle interaction properties. But there are a number of complicated and till unsolved problems, which doesn't allow to formulate the complete theory of formation and evolution of the objects in the Universe now. Let's give some examples of these up-to-date unsolved problems directly connected with nuclear astrophysics and nuclear interactions, which are followed from the existing to date nuclear physics problems: The insufficiency of experimental data of the nuclear reaction cross-sections at ultra low and astrophysical energies. This problem consists in the impossibility, at the modern stage of the development of experimental methods, to carry out direct measurements of the cross-sections of thermonuclear reactions in the earth's conditions for energies at which they are proceeding in the stars. We will stay at this problem more particularly, but now we will illustrate the main conceptions and representations generally using for the description of the thermonuclear reactions. The data of cross-sections or astrophysical S-factors of thermonuclear reactions, including radiative capture reactions and their analysis in the frame of different theoretical models, are the main source of information about nuclear processes taken place in the Sun and stars. The considerations of similar reactions are complicated by the fact that in many cases only theoretical predictions or extrapolation results can to make up deficient experimental information about characteristics of thermonuclear processes in stellar material at ultra low energies. The astrophysical S-factor, which determines the reaction cross-section, is the main characteristic of any thermonuclear reaction, i.e. the probability of reaction behavior at vanishing energies. It can be obtained experimentally, but it is generally possible for the majority of interacting nuclei, which are taken place in thermonuclear processes at the energy range above $100 keV \sim 1 MeV$, but for real astrophysical calculations, for example the developing of star evolution problem, the values of astrophysical S-factor, are required at the energy range about 0.1-100 keV, which corresponds to the temperatures in the star core on the order of $10^6 K \sim 10^9 K$. One of methods for obtaining the astrophysical S-factor at zero energy, i.e. the energy on the order of 1 keV and less, is the extrapolation of its values to lower energy range where it can be determined experimentally. It is the general way which is used, first of all, after carrying out the experimental measurements of cross-section of certain thermonuclear reaction at low energy range. The second and evidently most preferable method consists in theoretical calculations of the S-factor of some thermonuclear reaction on the basis of certain nuclear models. However, the analysis of all thermonuclear reactions in the frame of unified theoretical point of view is quite labor-consuming problem and later we will consider only photonuclear processes with g-quanta, specifically the radiative capture for certain light nuclei. The general sense of usage nuclear models and theoretical methods of calculation of thermonuclear reaction characteristics consists in the following. If the certain nuclear model correctly describes the experimental data of the astrophysical S-factor in that energy range where these data exist, for example $100keV \sim 1MeV$, then it is reasonably assume that this model will describe the form of the S-factor correctly at the most low energies (about 1 keV) too. This is the certain advantage of the approach stated above over the simple data extrapolation to zero energy, because the using model has, as a rule, the certain microscopic justification with a view to the general principles of nuclear physics and quantum mechanics. As for the model choice, one of these models, which we use in present calculations, is the potential cluster model (PCM) of the light atomic nuclei with the classification according to the Young schemes. The model, in the certain cases, contains the forbidden states (FS) for inter cluster interactions and in the simplest form gives a lot of possibilities for carrying out the similar other calculations. The PCM model is based on the assumption that the nuclei under consideration consist of two clusters. One may chose potential cluster model because of the fact that the probability of formation of isolated nucleon associations and their isolation in the majority of light atomic nuclei is relatively high. It is confirmed by numerous experimental data and theoretical results obtained over the last 5-6 decays. Thus, the one-channel potential cluster model could be a good approximation to the situation really existing in the atomic nucleus in many cases and for various light nuclei. A general approach that leads to the real results in the calculations of the astrophysical S-factor of the certain thermonuclear reaction with g-quanta, say, for the radiative capture reaction. For carrying out such calculations it is necessary to have the certain data and execute following steps:

1. Have at one's own disposal the experimental data of the differential cross sections or excitation functions σ_{exp} for the elastic scattering of the considered nuclear particles (for example $-p^2H$) at lowest energies known at the moment.

2. Carry out the phase shift analysis of these data or have the results of the phase shift analysis of similar data that were done earlier, i.e. to

know the phase shifts $\delta_l(E)$ of the elastic scattering dependent on the energy E. It is one of the major parts of the entire calculation procedure of the astrophysical S-factors in PCM with FS, since it allows to obtain the potentials of the inter cluster interaction.

3. Construct the interaction potentials V(r) (for example for p^2H system) according to the discovered phase shifts of scattering. This procedure is called as the potential description of the phase shifts of the elastic scattering in PCM with FS and it is necessary to carry out it at lowest energies.

4. It is possible to carry out the total cross-sections of the photo-decay process (for example $^3He + \gamma \rightarrow p +^2 H$) and the total cross-sections of the radiative capture ($p +^2 H \rightarrow^3 He + \gamma$) process connected with the previous by the principle of detailed balancing, if we have the inter-cluster potentials obtained in such a way, i.e. to obtain the total theoretical cross-sections S(E) of the photo nuclear reactions.

5. Then, it is possible to calculate the astrophysical S-factor of the thermonuclear reaction, for example $p +^2 H \rightarrow 3^H e + \gamma$, only if you have the total cross-sections of the radiative capture, i.e. the S(E) value as the function of energy E, at any lowest energies.

One may note that as of today the experimental measurements were done only for the astrophysical S-factor of the radiative p^2H capture down to 2.5keV, i.e. in the energy range which can be named as astrophysical. For all other nuclear systems taking part in the thermonuclear processes such measurements were thoroughly done only down to 50keV at the best, as it was done, for example, for the p^3H system. Schematically, all these steps can be represented in the next form:

$$\sigma_{exp} \rightarrow \delta_L(E) \rightarrow V(r) \rightarrow \sigma(E) \rightarrow S(E)$$

The way, stated above, is used in this review and identical for all photo nuclear reactions is independent of, for example, reaction energy or some other factors, and is the general at the consideration of any thermonuclear reaction with γ-quanta, if it is analyzed in the frame of potential cluster model with FS.

4.5.3   Sub-barrier fusion and selective resonant tunneling cross section

The cross section for sub-barrier fusion for light nuclei can be calculated using the selective resonant tunneling model assuming a complex square-well nuclear potential. The complex potential is assumed to describe the absorption inside the nuclear well. For the last 6o years Controlled nuclear fusion research has been concentrated much on deuteron-triton fusion because their fusion cross section is greater than that of deuteron-deuteron fusion by a factor of several hundred, although the Coulomb barrier for d+t is almost the same as that for d+d. The resonance of the d+t state near 100 keV is considered as the reason for such a large cross section. A simple square- well model with an imaginary part can be applied to describe the d+t nuclear interaction and the fusion reaction. It is interesting to notice that while the real part of the potential is mainly derived from the resonance energy, the imaginary part of the potential is determined by the Gamow factor at the energy of this resonance. The good agreement between the experimental data and the quantum-mechanics calculation suggests a selective resonant tunneling model, rather than the conventional compound nucleus model, because the penetrating particle will keep its memory of the phase factor of its wave function. The implication of this selective resonant tunneling model can further be explored for other light nuclei fusion. When a deuteron is injected to a triton, their relative motion can be described by a reduced radial wave function $\zeta(r)$, which is related to the solution of the Schrödinger equation, $\Phi(r,t)$ by

$$\Phi(r,t) = \frac{1}{\sqrt{4\pi}r}\zeta(r)\exp(-i\frac{E}{\hbar}t) \tag{4.47}$$

The Hamiltonian has an isotropic potential as shown in Figure 4.6 which is composed of a square well (r<a), and a Coulomb potential (r>a). Nuclear interaction would introduce a phase shift δ in the wave function, so, the cross section of the reaction σ_{re} in the low energy limit where only S-wave contributes may be related to the phase shift as

$$\sigma_{re} = \frac{\pi}{k^2}(1 - |\eta|^2) \tag{4.48}$$

where

$$\eta = e^{2i\delta_0} \tag{4.49}$$

and k is the wave number for the relative motion. For a complex nuclear potential , the phase shift δ_0 will also be a complex number. Hence, it is convenient to assume

$$cot(\delta_0) = W_r + iW_i \tag{4.50}$$

Then the tunneling probability (T) which can be given by

$$
\begin{aligned}
T &= 1 - |\eta|^2 \tag{4.51}\\
&= 1 - |e^{2i\delta_0}|^2 \tag{4.52}\\
&= 1 - \left| \frac{1 - icot\delta_0}{1 + icot\delta_0} \right|^2 \\
&= 1 - \left(\frac{1 - i(W_r + iW_i)}{1 + i(W_r + iW_i)} \right) \left(\frac{1 + i(W_r + iW_i)}{1 - i(W_r + iW_i)} \right) \\
&= 1 - \left(\frac{(1 + W_i) - iW_r}{(1 - W_i) + iW_r} \right) \left(\frac{(1 + W_i) + iW_r}{(1 - W_i) - iW_r} \right) \\
&= 1 - \left(\frac{(1 + W_i)^2 + W_r^2}{(1 - W_i)^2 + W_r^2} \right) \\
or,\ T &= -\frac{4W_i}{(1 - W_i)^2 + W_r^2} \tag{4.53}
\end{aligned}
$$

Connection between the Coulomb nuclear potential

The wave function inside the nuclear well (r<a) is determined by two parameters, the real and the imaginary part of the nuclear potential (U_{1r} and U_{1i}). The Coulomb wave function outside the nuclear well (r>a) is determined by two other parameters as well: the real and the imaginary part of the phase shift $(\delta_0)_r$ and $(\delta_0)_i$. A pair of convenient parameters, W_r and W_i, are introduced to make a linkage between the cross section and the nuclear well. Then, it is easy to discuss the resonance and the selectivity in damping. The connection of the wave function at the boundary (r=a) can be expressed by the logarithmic derivative of the wave function. In the square well, the dimensionless logarithmic derivative is

$$a \frac{[\sin(k_1 r)]'}{\sin(k_1 r)} \Big|_{r=a} = k_1 a \frac{\cos(k_1 a)}{\sin(k_1 a)} = k_1 a \cot(k_1 a). \tag{4.54}$$

In the Coulomb field, the dimensionless logarithmic derivative has been given by

$$\frac{a}{a_c} \left\{ \frac{1}{\chi^2} \cot(\delta_0) + 2 \left[\ln\left(\frac{2a}{a_c} \right) + 2A + y(ka_c) \right] \right\} \tag{4.55}$$

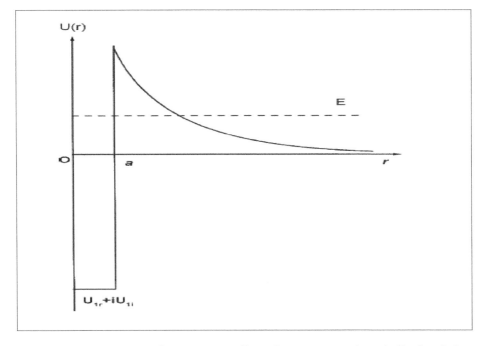

Figure 4.6: Schematics for square-well nuclear potential and Coulomb barrier.

Then

$$W_i = \chi^2 Im\left[\frac{a_c}{a}(k_1 a)\cot(k_1 a)\right] \tag{4.56}$$

$$= \chi^2\left\{\frac{a_c}{a}\frac{\gamma_i\sin(2\gamma_r) - \gamma_r\sinh(2\gamma_i)}{2[\sin^2(\gamma_r) + \sinh^2(\gamma_i)]}\right\} \tag{4.57}$$

$$W_r = \chi^2\left\{\frac{a_c}{a}\frac{\gamma_r\sin(2\gamma_r) + \gamma_i\sinh(2\gamma_i)}{2[\sin^2(\gamma_r) + \sinh^2(\gamma_i)]} - 2H\right\} \tag{4.58}$$

$$\chi^2 = \frac{1}{2\pi}\left\{\exp\left(\frac{2\pi}{ka_c}\right) - 1\right\} \tag{4.59}$$

$$k_1^2 = \frac{2\mu}{\hbar^2}[E - (U_{1r} + iU_{1i})] \tag{4.60}$$

$$k_{1i} = \frac{\mu}{k_{1r}\hbar^2}(-U_{1i}) \tag{4.61}$$

$$\gamma = \gamma_r + i\gamma_i \equiv k_{1r}a + ik_{1i}a \tag{4.62}$$

$$H = \left[\ln\left(\frac{2a}{a_c}\right) + 2A + y(ka_c)\right] \tag{4.63}$$

$$k^2 = \frac{2\mu E}{\hbar^2} \tag{4.64}$$

$$a = a_0(A_1^{1/3} + A_2^{1/3}), \ [a_0 = 1.12 \times 10^{-13} cm] \tag{4.65}$$

$$a_c = \frac{\hbar^2}{Z_1 Z_2 \mu e^2} \tag{4.66}$$

$$y(x) = \frac{1}{x^2}\sum_{j=1}^{\infty}\frac{1}{j(j^2 + x^{-2})} \tag{4.67}$$

In the above relations k is the wave number outside the nuclear well, a_c is the Coulomb unit of length, and $A = 0.577$ is Euler's constant, $y(ka_c)$ is related to the logarithmic derivative of Γ function given by the Eq. (4.66), δ_0 is the complex phase shift of the wave function due to the nuclear interaction, $1/\chi^2$ is the Gamow penetration factor of the Mott form, a is the radius of the nuclear well, A_1 and A_2 are the mass numbers of the colliding nuclei, $Z_1 e$ and $Z_2 e$ are their electrical charges, K_1 is the wave number inside the nuclear well. We have from Eq. (4.46)

$$\sigma_{re} = \frac{\pi}{k^2}\left(-\frac{4W_i}{(1 - W_i)^2 + W_r^2}\right) \tag{4.68}$$

It can be seen from (4.49) that σ_{re} reaches maximum for

$$W_r = 0$$
$$W_i = -1 \tag{4.69}$$

It is evident that $W_r = 0$ corresponds to the condition for resonance, i.e.,

$$Re(\delta_0) = \frac{n\pi}{2} \tag{4.70}$$

where n is an odd integer. On the other hand, W_i is connected to the imaginary part of the nuclear potential, U_{1i}. When $U_{1i} = 0, W_{1i} = 0$. It simply means that the fusion cross section is zero, if there is no absorption. However, if $U_{1i} \to -\infty$, $|W_i| \gg 1$. It means that the cross section of the fusion reaction is then proportional to $1/|W_i| \ll 1$, also when the absorption is very strong. In other words, there must be a suitable value of U_{1i} in between, which makes the fusion cross section maximized at the resonance. This is the value of U_{1i} which makes $W_i = -1$ at $W_r = 0$. This can be understood if we notice that absorption acts like damping in a resonance. The energy absorbed by a damping mechanism is proportional to the product of the damping coefficient and the square of the amplitude of the oscillation. When the damping coefficient is zero, the energy absorbed by damping mechanism is zero even if the resonance develops fully. On the other hand, when the damping coefficient is too large, the damping mechanism will kill the resonance before it is fully developed. Thus, the energy absorbed by the damping mechanism is still very small. Hence, there must be a suitable damping which makes the absorbed energy maximized. Similarly, the fusion cross section is proportional to the product of U_{1i} and the square of the amplitude of the wave function inside the nuclear well; therefore, there should be a suitable damping U_{1i} to make the fusion cross section maximized. We may call it matching damping. Consequently, one may ask the question if such a matching damping manifests itself in a nuclear resonant process.

Experimental evidence

In experiment, at the resonance energy the resonant state with the matching damping will have the largest tunneling current; hence, it should be observed first. This may be checked directly through the experimental data. The famous d+t fusion process is the best candidate for this purpose, because it has a well-known resonance at the energy of 114 keV. If we assume that at this resonant energy not only $W_r = 0$, but also $W_i = -1$ to maximize the tunneling current; then, the theoretical prediction for the fusion cross section due to the S wave should be

$$\sigma_{reso} = \frac{\pi}{k^2} = 4.74 barn. \tag{4.71}$$

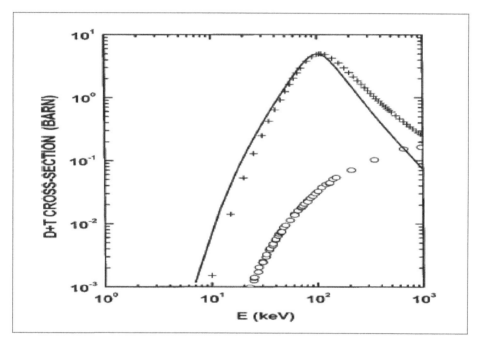

Figure 4.7: Fusion cross section: experimental data for d+t fusion (+); selective resonant tunneling calculation for d+t fusion (solid line); and experimental data for d+d fusion (o). [Source: PRC 61, 024610(2000)]

The experimental value for the fusion cross section due to all the partial waves is

$$\sigma_{expt} = 4.98 barn. \tag{4.72}$$

Moreover, based on the assumption of Eq.(4.53), we may calculate the nuclear potential under the square-well assumption. The real part (U_{1r}) and imaginary part (U_{1i}) of the nuclear potential are obtained as

$$U_{1r} = -41.4 MeV$$
$$U_{1r} = -123.0 keV \tag{4.73}$$

Using these parameters for the nuclear well, we may further calculate the phase shift as a function of energy $\delta_0(E)$ as well as the cross section $\sigma_{re}(E)$ as a function of energy E for the S-wave. Figure 4.7 shows the result of calculation which are obtained following Eq.(4.53) using the nuclear well parameters of Eq. (4.58) for d+t and d+d respectively. The results are in good agreement in the low-energy side. The contribution from the p wave may further improve the agreement on the high-energy side.

Selective resonant tunneling

It is interesting to discuss the tunneling probability T in Eq.(4.53). The resonant feature is clearly shown by the dependence on W_r (Figure 4.8). The tunneling probability will reach its peak at $W_r = 0$, and the width of this peak is determined by $|W_i - 1|$. When $W_i = -1$, $T = 1$. If W_i is greater or less than (-1); then, the peak value of T is always less than 1 (see dashed lines and dotted lines in Figure 4.8). This will generate a selective feature for resonant tunneling phenomenon. As we may expect, W_r (i.e., mainly the real part of the phase shift varies with the incident energy of the projectile. However, W_i varies with the lifetime of the state which is composed of the tunneling projectile and the target. When the incident energy is in resonance with the energy level of the composed state, the resonant tunneling happens $(W_r = 0)$. However, if the lifetime of this composed state does not make $W_i = -1$; then, the tunneling probability is still low even if at this resonant energy resonant energy. Thus, if there are more than one states with different lifetimes at the same energy level; then, the resonant tunneling process may generate only a few states which have the right lifetime to make $W_i \simeq -1$. We may call it selective resonant tunneling. Now the question is how sharp is this selectivity. At the resonance, W_i will be given by Eq. (4.57). From expressions (4.57), (4.60) and (4.61), we

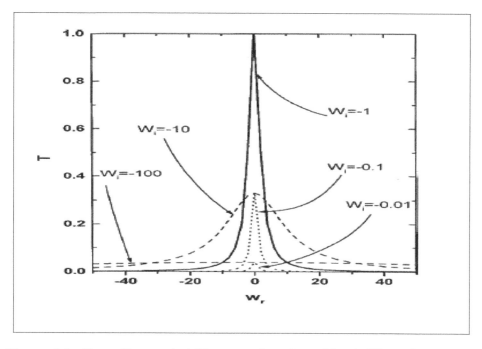

Figure 4.8: Tunneling probability as a function of both W_r and W_i. The solid line is for $W_i = -1$; the dashed line is for $W_i \ll -1$; the dotted line is for $0 > W_i > -1$. [Source: PRC 61, 024610(2000)]

may observe the dependence of W_i on U_{1i}. When $U_{1i} = 0$ (no absorption), $\gamma_i \equiv k_{1i}a = 0$; then, $W_i = 0$ and $T = 0$. On the other hand, if $|U_{1i}|$ is very large and it makes $|\gamma_i| \equiv |k_{1i}a| \gg 1$; then, $|W_i$ rises quickly with $(k_{1i}a)$ due to large factor χ^2 in Eq. (4.57). However, there is a suitable value for $|U_{1i}|$ to make $T = 1$ at resonance. It is possible to make $W_i = -1$ if $|U_{1i}|$ is a small number of the order of $\frac{1}{\chi^2}$. Which we may call the matching damping value determined by the Gamow factor χ^{-2}. When the damping deviates from this matching damping value, the tunneling probability T will approach zero quickly due to the large factorχ^2 in front of Eq. (4.57). Usually, for the low-energy tunneling ($k^2 \to 0$), χ^2 is a very large number due to the exponential factor in Eq.(4.59). Hence, the selectivity on U_{1i} would be very sharp. Tunneling probability at resonance would vary from 0 to 1 when U_{1i} changes from 0 to $-\chi^{-2}$. Tunneling probability would drop quickly from 1 to $O(1/\chi)$ when U_{1i} changes from $-\frac{1}{\chi^2}$ to $-\frac{1}{\chi}$. This is a very sharp selectivity on damping.

Selective resonant tunneling versus compound nucleus model

The selective resonant tunneling model means that nuclear resonance selects not only the frequency (energy level), but also the damping (nuclear reaction). The selectivity becomes very sharp, when the resonance happens in a low energy sub-barrier tunneling. Thus, the neutron-emission reaction is suppressed in such selective resonant tunneling processes. The compound nucleus model may not be applied to light nuclei fusion, because the penetrating particle may still remember its phase factor of the wave function, while the compound nucleus model assumes that the penetrating particle loses memory of its history. In the compound nucleus model, the nuclear reaction is divided into two steps: penetrating first, then decaying. In selective resonant tunneling, the tunneling probability depends on the lifetime of decay. The tunneling process is completed in one single step. The surprisingly good agreement between the calculated cross section and experimental value for d+t sub-barrier fusion is strong evidence showing that, the tunneling process is a single step process. The discovery of the nuclear halo state is another strong evidence showing that even if inside the strongly interacting nuclear well region, the nucleon may still keep its own feature without losing its memory of the wave function. The Breit-Wigner formalism for the resonant interaction requires two parameters for each resonance: the energy and the width for the resonance. However, the selective resonant tunneling model for sub-barrier fusion requires only one parameter-the energy of the resonance; the width of the resonance is then determined by the Gamow factor $[U_{1i} \simeq -O(\chi^{-2})]$. In the calculation of the curve for the d+t cross section in Figure 4.7, no input from the experiment for width has been used, rather $W_i = -1$ has been assumed for the maximum tunneling, giving the result of the selective resonant tunneling. As pointed by Balantekin the fusion of two nuclei at very low energies are not only of central importance for stellar energy production and nucleosynthesis, but also provide new insights into reaction dynamics and nuclear structure.

Astrophysical S-function

Selective resonant tunneling model also provides a convenient way to calculate astrophysical S-function defined by Bosch[42] as

$$S(E) = \exp\left(\frac{2\pi}{ka_c}\right) E\sigma(E) \tag{4.74}$$

---

[42]Bosch, H. S. and Hale, G. M.: Nucl. Fusion 32, 611(1992). 15.

$$= \frac{\pi^2 \hbar^2}{\mu} \left\{ \frac{\exp\left(\frac{2\pi}{ka_c}\right)}{\exp\left(\frac{2\pi}{ka_c}\right) - 1} \right\} \left\{ \frac{-4\omega_i}{\omega_r^2 + (\omega_i - \chi^{-2})^2} \right\} \quad (4.75)$$

$$= (\frac{\pi}{k^2}) \underbrace{\left\{ \frac{2\pi}{\exp(\frac{2\pi}{ka_c}) - 1} \right\}}_{} \underbrace{\left\{ \frac{-4\omega_i}{\omega_r^2 + (\omega_i - \chi^{-2})^2} \right\}}_{} \quad (4.76)$$

In Eq. (4.76) the first factor is a geometric factor, second factor is the Gamow factor and the third factor is the S-factor and the complex variable ω is connected to W (also complex) via

$$\omega = \omega_r + i\omega_i = W/\chi^2 = (W_r + iW_i)/\chi^2 \quad (4.77)$$

and

$$\sigma = \frac{\pi}{k^2} \frac{1}{\chi^2} \left\{ \frac{-4\omega_i}{\omega_r^2 + (\omega_i - \chi^{-2})^2} \right\} \quad (4.78)$$

Thus S(E) can be extracted as [43]

$$S(E) = \left\{ \frac{-4\omega_i}{\omega_r^2 + (\omega_i - \chi^{-2})^2} \right\} \quad (4.79)$$

Eqs. (4.68) and (4.78) are different from Breit-Wigner's formula for resonance cross section written as

$$\sigma \propto \frac{1}{(E - E_r)^2 + (\frac{\Gamma}{2})^2} \quad (4.80)$$

In Eqs. (4.68) or (4.78) there is no Taylor expansion to obtain in terms of $(E - E_r)^2$, hence Eqs. (4.68) or (4.78) are more accurate in the wide range of energy for the practical uses.

4.6 Weak-Interaction Rates

This section presents a discussion on recent progress in electron capture and beta decay rates via large scale shell model calculations and for neutrino induced reactions in general. Decay studies for heavy nuclei, where shell model calculations are presently not feasible and large scale QRPA calculations are employed, have been addressed.

---

[43] Nucl. Fusion 48 (2008) 125003

4.6.1   Electron Capture and β-Decay

For densities $\rho \leq 10^{11} g/cm^3$, stellar weak-interaction processes are dominated by Gamow-Teller (GT) and, if applicable, by Fermi transitions. (One distinguishes between GT_+ transitions, where a proton is changed into a neutron, and GT_- transitions, where a neutron is changed into a proton. Electron capture is sensitive to the GT_+ strength, while ordinary β decay depends on the GT_- distribution.) At higher densities forbidden transitions have to be included as well. To understand the requirements for the nuclear models to describe these processes (mainly electron capture), it is quite useful to recognize that electron capture is governed by two energy scales: the electron chemical potential (or Fermi energy) μ_e, which grows like $\rho^{1/3}$, and the nuclear Q-value. As is sketched in Figure 4.9, μ_e grows much faster than the Q values of the abundant nuclei. We can conclude that at low densities, where one has $\mu_e \sim Q$ (i.e. during late hydrostatic burning), the capture rate will be very sensitive to the phase space and requires an accurate as possible description of the detailed GT_+ distribution of the nuclei involved. Furthermore, the finite temperature in the star requires the implicit consideration of capture on excited nuclear states, for which the GT distribution can be different than for the ground state. As we will demonstrate below, modern shell model calculations are capable to describe GT_+ rather well and are therefore the appropriate tool to calculate the weak-interaction rates for those nuclei ($A \sim 50 - 65$) which are relevant at such densities. At higher densities, when μ_e is sufficiently larger than the respective nuclear Q values, the capture rate becomes less sensitive to the detailed GT_+ distribution and is mainly only dependent on the total GT strength. Thus, less sophisticated nuclear models might be sufficient. However, one is facing a nuclear structure problem which has been overcome only very recently. We come back to it below, after we have discussed the calculations of weak-interaction rates within the shell model and their implications to pre-supernova models. The general formalism to calculate weak interaction rates for stellar environment has been given by Fuller, Fowler and Newman (FFN)[44] who also estimated the stellar electron capture and beta-decay rates systematically for nuclei in the mass range A = 20 - 60 based on the independent particle model and on data, whenever available. In recent years this pioneering and seminal work has been replaced by rates based on large-scale shell model calculations. Similar calculations for pf-shell nuclei had to wait until significant progress in shell model diagonalization, mainly due to Etienne Caurier, allowed calculations in either the full pf shell or at such a truncation level that the GT

---

[44]G.M. Fuller, W.A. Fowler, M.J. Newman: Astrophys. J. 293, 1 (1985)

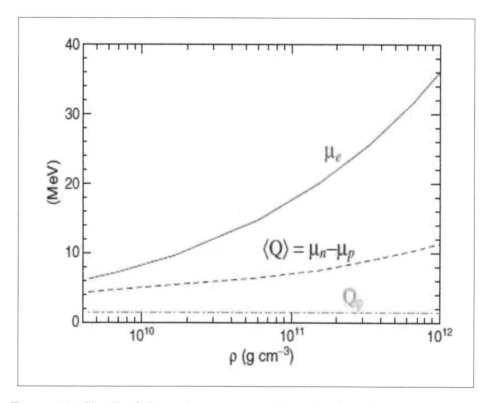

Figure 4.9: Sketch of the various energy scales related to electron capture on protons and nuclei as a function of density during a supernova core collapse simulation. Shown are the chemical potential / Fermi energy of electrons, the Q-values for electron capture on free protons (constant) and the average Q-value for electron capture on nuclei for the given composition at each density.

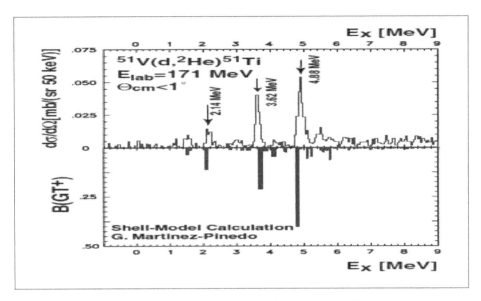

Figure 4.10: Comparison of the measured $^{51}V(d,^2He)^{51}$Ti cross section at forward angles (which is proportional to the GT_+ strength) with the shell model GT distribution in ^{51}V [Source: Phys. Rev. C 68, 031303 (2003)].

distributions were virtually converged. It has been demonstrated by Courier et al [45] that the shell model reproduces all measured GT_+ distributions very well and gives a very reasonable account of the experimentally known GT_- distributions. Further, the lifetimes of the nuclei and the spectroscopy at low energies is simultaneously also described well. Charge-exchange measurements using the $(d,^2He)$ reaction at intermediate energies allow now for an experimental determination of the GT_+ strength distribution with an energy resolution of about 150 keV. Figure 4.10 compares the experimental GT_+ strength for ^{51}V, measured at the KVI in Groningen[46], with shell model predictions. It can be concluded that modern shell model approaches have the necessary predictive power to reliably estimate stellar weak interaction rates. Such rates have been calculated by K. Langanke et al[47] for more than 100 nuclei in the mass range A = 45-65. The rates have been calculated for the same temperature and density grid as the standard FFN compila-

---

[45]see E. Caurier, K. Langanke, G. Martinez-Pinedo, F. Nowacki: Nucl. Phys. A 653, 439 (1999)
[46]C. Bäumer et al.: Phys. Rev. C 68, 031303 (2003)
[47]K. Langanke, G. Martinez-Pinedo: At. Data Nucl. Data Tables 79, 1 (2001)

tions[48]. An electronic table of the rates is available[49]. Importantly one finds that the shell model electron capture rates are systematically smaller than the FFN rates. The difference is particularly large for capture on odd-odd nuclei which have been previously assumed to dominate electron capture in the early stage of the collapse.

4.6.2 Neutrino-Induced Reactions

While elastic scattering of neutrinos on nuclei is important for neutrino opacities to describe the leakage and escape of neutrinos, neutrino-induced reactions on nuclei can also play a role at high densities during the stellar collapse and explosion phase in causing composition changes. It is to be noted that during the collapse only ν_e neutrinos are present. Thus, charged-current reactions $A(\nu_e, e^-)A'$ are strongly blocked by the large electron chemical potential. Inelastic neutrino scattering on nuclei can compete with ν_+e^- scattering at higher neutrino energies $E_\nu \geq 20MeV$. At such energies forbidden transitions can contribute noticeably to the cross sections. Finite-temperature effects play an important role for inelastic $\nu+A$ scattering below $E_\nu \leq 10MeV$. This comes about as nuclear states, which are connected to the ground state and low-lying excited states by modestly strong GT transitions and increased phase space, get thermally excited. As a consequence the cross sections are significantly increased for low neutrino energies at finite temperature and might be comparable to inelastic $\nu_e + e^-$ scattering. Thus, inelastic neutrino-nucleus scattering, which is so far neglected in collapse simulations, should be implemented in such studies. This is in particular motivated by the fact that it has been demonstrated that electron capture on nuclei dominates during the collapse and that this mode generates significantly less energetic neutrinos than considered previously (see below). Examples for inelastic neutrino-nucleus cross sections are shown in Figure 4.11. A reliable estimate for these cross sections requires the knowledge of the GT_0 strength. Shell model predictions imply that the GT_0 centroid resides at excitation energies around 10 MeV and is independent of the pairing structure of the ground state. Finite temperature effects become unimportant for stellar inelastic neutrino-nucleus cross sections once the neutrino energy is large enough to reach the GT_0 centroid, i.e. for $E_\nu \geq 10MeV$.

---

[48]G.M. Fuller, W.A. Fowler, M.J. Newman: Astro. phys. J. 252, 715 (1982)

[49]K. Langanke, G. Martinez-Pinedo: At. Data Nucl. Data Tables 79, 1 (2001)

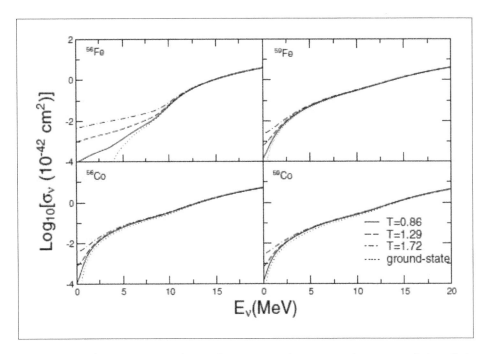

Figure 4.11: Cross sections for inelastic neutrino scattering on nuclei at finite temperature. The temperatures are given in MeV[Ref: Phys. Lett. B 529 (2002) 19].

Solved Problems

1. Saha equation may be written as,

$$N_{xy} = \frac{N_x N_y}{1 + \delta_{xy}} \left(\frac{2\pi}{\mu kT}\right)^{3/2} \hbar^3 \omega exp(-\frac{E_R}{kT})$$

Explain the meaning of the terms $N_{xy}, N_x, N_y, \delta_{xy}, \mu, \omega, E_R$ in the above expression.

Solution a) In the above expression N_{xy} is the number of nuclei in the resonant state at energy E_R (for $x + y \to p$), N_x is the number of nuclei of the type x, N_y is the number of nuclei of the type y, δ_{xy} is the Kronecker delta having value 1(for x, y identical) or 0 (for x, y non-identical), $\mu = \frac{m_x m_y}{m_x + m_y}$ is the reduced mass of the system, E_R is the energy of the resonant state, k is the Boltzmann constant, T is the absolute temperature, $\omega = \frac{2j+1}{(2j_x+1)(2j_y+1)}$ is a statistical factor determined by the spins of the nuclei x, y.

2. Calculate the effective burning energy of two alpha-particles in a stellar gas of temperature (a) $10^7 K$, (b) $5 \times 10^7 K$ and (c)$10^8 K$.

Solution The effective buring energy is defined as

$$E_{eb} = 1.22(Z_1^2 Z_2^2 \mu T_6^2)^{1/3} keV,$$

where T_6 is in in unit of $10^6 K$.
a) $\mu = \frac{A_1 A_2}{A_1 + A_2} = \frac{A_\alpha^2}{2A_\alpha} = 2, T_6 = 10,$

$$E_{eb} = 1.22(2^2 \times 2^2 \times 10^2 \times 2 \times 10^2)^{1/3} keV = 68.4 keV$$

b) $T_6 = 5 \times 10,$

$$E_{eb} = (25)^{1/3} \times 68.4 keV = 200 keV$$

c)$T_6 = 10 \times 10,$

$$E_{eb} = (100)^{1/3} \times 68.4 keV = 317 keV$$

Exercise

1. Using Saha equation estimate the ratio $\frac{^8 Be}{^4 He}$ in a gas of ^{4}He atoms at temperature $10^8 K$ with a density of $3 \times 10^8 gm/cm^3$, $k = 1.38 \times 10^{-23} JK^{-1}$, Q-value for $2\alpha \to^8 Be$ is -92keV which corresponds to the resonance state at energy $E_R = 92$keV.

2. Sketch the variation of fusion rate as a function of particle energy for the fusion of two charged nuclei in a gas of particles which follow a Maxwell- Boltznmann type distribution. On the same sketch, show the variation of probability of barrier penetration as a function of energy. Using these sketches, explain the orgin of the 'Gamow peak'. Also extract from the expression for fusion rate, expression for the effective burning energy $E_{eb} = 1.22(Z_1^2 Z_2^2 \mu T_6^2)^{1/3} keV$ and the energy width $(\Gamma = \Delta E)$ of the Gamow peak.

3. Calculate the effective burning energy for $^{12}C(\alpha, \gamma)^{16}O$ at $T = 2 \times 10^8 K$ using the expression $E_{max} = (bkT/2)^{2/3}$, where $b = \pi \alpha Z_1 Z_2 \sqrt{2\mu c^2} = 0.990 Z_2 Z_2 \sqrt{\mu} \sqrt{MeV}$, $\mu = \frac{A_1 A_2}{A_1 + A_2}$ in a.m.u. [Ans: $E_{max} = 0.3 MeV$]

4. Calculate the effective burning energy and width of Gamow peak for an ensemble of ^{4}He nuclei at a temperature of $T = 10^9$K. Note Q-value $\alpha + \alpha \rightarrow^8 Be$ is -92 keV, k = 1.384 $\times 10^{-23}$ JK^{-1}, $\hbar = 1.05 \times 10^{-34}$ Js, $m_u = 931.5 MeV/c^2 \simeq 1.66 \times 10^{-27}$kg,$T_6$ = temperature in units of 10^6K.

5. Calculate the classical cross-section of (i) a proton on ^{12}C, (ii) ^{12}C on ^{12}C [Hint: $\sigma_{classical} = \pi(R_p + R_t)^2$]

6. Calculate the de Broglie wavelength of a proton fired at a ^{12}C target at an energy of (i) 100 keV and (ii) 1 MeV. Use de calculated de Broglie wavelength to re-calculate cross section for a beam of protons at an energy of 5 MeV onto a ^{12}C target at rest. [Hint:$\lambda_{dB} = \left(\frac{M_p + M_t}{M_t}\right) \times \left(\frac{\hbar}{\sqrt{M_p E_{lab}}}\right)$, $\sigma = \frac{\lambda_{dB}^2}{4\pi}$]

7. Calculate the Sommerfeld parameter, ν, using the expression

$$\nu = \frac{31.29 Z_1 Z_2 \sqrt{\mu/E}}{2\pi},$$

where μ is the reduced mass of the system in a.m.u. and E is the energy in keV of the collision in the centre of mass frame. Using the value of the cross-section calculated in the previous problem, estimate the value of the astrophysical S-factor for the compound reaction ^{12}C +p$\rightarrow^{13}$N at a proton energy of 5MeV in the CM frame.

8. Sketch the variation of cross-section with particle energy for the fusion of two charged nuclei in a gas of particles with velocities which follow a Maxwell-Boltzmann type distribution. Explain the origin of the 'Gamow Peak' using this diagram.

9. Given that
$$E_G = 1.2(Z_1^2 Z_2^2 \mu T_6^2)^{1/3} keV,$$

where E_G is the Gamow peak energy, calculate the effective burning energy of the $^{12}C + p \rightarrow^{13} N$ reaction at a temperature of 10^{10}K. Note, $\mu = \frac{A_1 A_2}{A_1 + A_2}$ is the reduced mass in atomic mass unit.

10. Given the expression

$$\frac{\sigma v_{cd}}{\sigma v_{ab}} = \frac{(2J_a + 1)(2J_b + 1)}{(2J_c + 1)(2J_d + 1)} \left(\frac{1 + \delta_{cd}}{1 + \delta_{ab}}\right) \left(\frac{\mu_{ab}}{\mu_{cd}}\right)^{3/2} \exp\left(-\frac{Q}{kT}\right)$$

Briefly explain the meaning of each term. Discuss the significance of this expression in the experimental determination of reaction rates.

11. Calculate the ratio of reaction rates per particle pair for the reaction $^7Li(p, \alpha)\alpha$ at T=10^9 and 10^{10}K. The ground state spins of the proton and ^{7}Li nuclei are $\frac{1}{2}^+$ and $\frac{3}{2}^-$ respectively. [Hint: Q-value and Boltzmann constant will be used]

12. Derive the relationship between the reaction rate per particle pair, $<\sigma v>$ and the total reaction rate per unit volume per unit time.

13. Calculate the probability for α-particle tunneling and fusing with a ^{16}O nucleus at a centre of mass energy of 0.1, 1 and 10 MeV.

14. Calculate the energy of the emitted photon in the direct capture reaction $^3He(\alpha, \gamma)^7Be$ to the $\frac{1}{2}^-$ excited state at 429 keV in ^{7}Be. Assuming ^{3}He at rest and the α-particle energy of 300 keV.

15. i) What is the energy width of a state which gamma-decays with a lifetime of 5×10^{-12}s? (ii) If this state can now α decay with a partial width of 10 eV, what is the new decay lifetime of the state? (iii) The width of the ground state resonance in ^{8}Be is 6.8 eV, what is the lifetime of ^{8}Be? Why does this resonance not have a γ-decay branch?

16. Explain the terms resonance and subthreshold resonance. Give an example of an important subthreshold resonance in a nuclear astrophysics reaction.

17. Discuss the effects of nuclear structure in the reactions involved in helium burning from the formation of ^{12}C up to ^{20}Ne. Comment on the effect of subthreshold and un-natural parity resonant states.

Refences

Quoted in the text:

- W.A. Fowler, G. E. Caughlan, B. A. Zimmermann: Ann. Rev. Astron. Astro- phys. 5, 525 (1967)

- D.D. Clayton: Principles of Stellar Evolution and Nucleosynthesis, (Univ. of Chicago Press 1968, 1983)

- C. Rolfs, W.S. Rodney: Cauldrons in the Cosmos, (University of Chicago Press, Chicago 1988)

- Arnett, W.D.: Nucleosynthesis and Supernovae, (Princeton Univ. Press 1996)

- F. K$\ddot{}$·appeler, F.-K. Thielemann, M. Wiescher: Ann. Rev. Nucl. Part. Sci. 48, 175 (1998)

- K. Wisshak, F. Voss, F. K$\ddot{}$·appeler, I. Kerzakov: Nucl. Phys. A 621, 270c (1997)

- M. Wiescher, V. Harms, J. G$\ddot{}$·orres, F.-K. Thielemann, L.J. Rybarcyk: Astrophys. J. 316, 162 (1987)

- D. Thomas, D.N. Schramm, K.A. Olive, B.D. Fields: Astrophys. J. 406, 569 (1993)

- T. Rauscher, J.H. Applegate, J.J. Cowan, F.-K. Thielemann, M. Wiescher: Astrophys. J. 429, 499 (1994)

- H. Schatz, A. Aprahamian, J. G$\ddot{}$·orres, M. Wiescher, T. Rauscher, J.F. Rembges, F.-K. Thielemann, B. Pfeiffer, P. M$\ddot{}$·oller, K.-L. Kratz, H. Herndl, B.A. Brown, H. Rebel: Phys. Rep. 294, 168 (1998)

- F.-K. Thielemann, M. Arnould, J. W.Truran: In: Advances in Nuclear Astrophysics, ed. E. Vangioni-Flam (Editions Frontiere, Gif sur Yvette 1987) p. 525

- J.J. Cowan et al.: Phys. Rep. 208, 267 (1991)

- T. Rauscher et al.: Phys. Rev. C 57, 2031 (1997)

- T. Rauscher, F.-K. Thielemann: At. Data Nucl. Data Tables 75, 1 (2000); 79, 47 (2001)

- D. Mocelj, T. Rauscher, G. Martinez-Pinedo, Y. Alhassid: In: Capture Gamma Ray Spectroscopy and Related Topics. (World Scientific, Singapore 2003) p. 781

- E.E. Salpeter, H.M. van Horn: Astrophys. J. 155, 183 (1969)

- L. Wilets, B.G. Giraud, M.J. Watrous, J.J. Rehr: Astrophys. J. 530, 504 (2000)

- . G. Chabrier, E. Schatzman: The Equation of State in Astrophysics, IAU Coll. 147, (Cambridge Univ. Press 1994)

- S. Schramm, S.E. Koonin: Astrophys. J. 365, 296 (1990)

- M. Beard, M. Wiescher: Rev. Mex. Fis. Suppl. 49, 139 (2003),

- P. Mohr, K.Vogt, M. Babilon et al.: Phys. Lett. B 488, 127 (2000)

- G.M. Fuller, W.A. Fowler, M.J. Newman: Astrophys. J. Suppl. 42, 447 (1980)

- S.W. Bruenn, W.C. Haxton: Astr. J. 376, 678 (1991)

- S.E. Woosley, D. Hartmann, R.D. Hoffman, W.C. Haxton: Astrophys. J. 356, 272 (1990)

- H.V. Klapdor, J. Metzinger, T. Oda: At. Data Nucl. Data Tables 31, 81 (1984)

- K. Takahashi, K. Yokoi: At. Data Nucl. Data Tables 36, 375 (1988)

- J. Görres, M. Wiescher, F.-K. Thielemann: Phys. Rev. C 51, 392 (1995)

- W.H. Press, B.P. Flannery, S.A. Teukolsky, W.T. Vetterling: Numerical Recipes, (C ambridge University Press 1986)

- W.R. Hix, F.-K. Thielemann; J. Comp. Appl. Math. (JCAM) 109, 321 (1999)

- G. Audi et al.: Nucl. Phys. A 624, 1 (1997)

- Y. Aboussir et al.: At. Data Nucl. Data Tables 61, 127 (1995)

- Arnett, W.D.: Nucleosynthesis and Supernovae, (Princeton Univ. Press 1996)

- S.E. Woosley, T.A. Weaver: Astrophys. J. Suppl. 101, 181 (1995)

- F. Käppeler, F.-K. Thielemann, M. Wiescher: Ann. Rev. Nucl. Part. Sci. 48, 175 (1998)

- V. Trimble: Astron. Astrophys. Rev. 3, 1 (1991)

- F.-K. Thielemann, D. Argast, F. Brachwitz et al.: Nucl. Phys. A 718, 139 (2003)

- K.H. Bockhoff, A.D. Carlson, O.A. Wasson, J.A. Harvey, D.C. Larson: Nucl. Sci. Eng. 106, 192 (1990)

- C. Rolfs, W.S. Rodney: Cauldrons in the Cosmos, (University of Chicago Press, Chicago 1988)

- R.V. Wagoner: Astrophys. J. Suppl. 18, 247 (1969)

- T. Rauscher, F.-K. Thielemann: At. Data Nucl. Data Tables 75, 1 (2000); 79, 47 (2001)

- J.A. Cizewski et al.: "Single-Particle Structure of Neutron-Rich Nuclei", In: Proceedings of the International Conference on the Labyrinth in Nuclear Structure, (AIP in press)

- M. Henchek et al.: Phys. Rev. 50, 2219 (1994)

- T. Delbar et al.: Phys. Rev. C 48, 3088 (1993)

- X. Gu et al.: Phys. Lett. B 343, 31 (1995)

- D.W. Bardayan et al.: Phys. Rev. C 63, 065802 (2001)

- T. Motobayashi et al.: Phys. Rev. Lett. 73, 2680-2683 (1994) Nuclear Astrophysics and Nuclei Far from Stability 463

- W. Hauser, H. Feshbach: Phys. Rev. A 87, 366 (1952)

- J.W. Truran, A.G.W. Cameron, A. Gilbert: Can. J. Phys. 44, 563 (1966)

- G. Michaud, W.A. Fowler: Phys. Rev. C 2, 2041 (1970)

- M. Arnould: Astr. and Astrophys. 19, 92 (1972)

Chapter 5

Explosive Burning Processes

5.1 Introduction

Processes like hydrogen and helium burning, where the stellar energy loss is dominated by the photon luminosity, choose temperatures with energy generation rates equal to the radiation losses. For the later burning stages neutrino losses play the dominant role among cooling processes and the burning timescales are determined by temperatures where neutrino losses are equal to the energy generation rate. Explosive events are determined by hydrodynamics, causing different temperatures and timescales for the burning of available fuel. We can generalize the question by defining a burning timescale according to Eq. (4.15) for the destruction of the major fuel nuclei

$$\tau_i = \frac{Y_i}{\dot{Y}_i}. \tag{5.1}$$

These timescales for the fuels ($i \in H, He, C, Ne, O, Si$) are determined by temperature dependent major destruction reactions. And when these dominating reactions are decay or photo disintegration processes, the timescale do not depend on density. This applies to Ne- and Si- burning, which are dominated by (γ, α) destructions of ^{20}Ne and ^{28}Si. Here the timescales are only determined by the burning temperatures. If the dominating reaction is (i) a two-particle or (ii) a three-particle fusion reaction, they have an inverse (i) linear or (ii) quadratic density dependence. Thus, in the burning stages which involve a fusion process, the density dependence is linear, with the exception of He-burning, where it is quadratic due to the triple-alpha reaction. The temperature dependences are typically exponential, due to the functional form of the corresponding $N_A < \sigma v >$ expressions. These burning

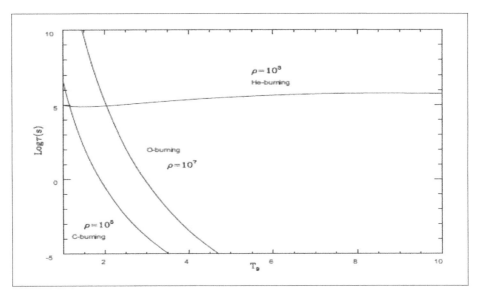

Figure 5.1: Burning timescales for fuel destruction of He-, C-, and O-burning as a function of temperature assuming 100% fuel mass fraction. The factor $N_{i,i}^i = \frac{N_i}{N_i!}$ cancels for the destruction of identical particles by fusion reactions, as $N_i = 2$. For He-burning the destruction of three identical particles has to be considered, which changes the leading factor $N_{i,i,i}^i$ to $1/2$. The density-dependent burning timescales are labeled with the chosen typical density (typical densities for He, C and O are respectively $10^3, 10^5$ and 10^7 in unit of gm/cm^3). They scale linearly for C- and O-burning and quadratically for He-burning. Notice that the almost constant He-burning timescale beyond $T_9 = 1$ has the effect that efficient destruction on explosive timescales can only be attained for high densities.

timescales as a function of temperature have been plotted in Figures 5.1 and 5.2, assuming a fuel mass fraction of 1 (i.e. 100%). The curves for (also) density dependent burning processes are labeled with a typical density. Typical explosive burning timescales of the order of seconds (e.g. in supernovae), one requires temperatures in excess of 4×10^9K (Si-burning), 3.3×10^9 K (O-burning), 2.1×10^9K (Ne-burning), and 1.9×10^9 K (C-burning) for burning essential parts of the fuel. Beyond 10^9 K He-burning is determined by an almost constant burning timescale. One may see that essential destruction on a time scale of 1s is only possible for densities $\rho > 10^5 gcm^{-3}$. This is usually not encountered in He-shells of massive stars. In a similar way explosive H-burning is not of relevance for massive stars, but important

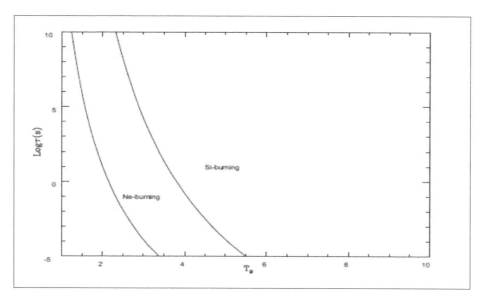

Figure 5.2: Burning timescales for fuel destruction of Ne- and Si-burning as a function of temperature. These are burning phases initiated by photo disintegrations and therefore not density-dependent.

for explosive burning in accreted H-envelopes in binary stellar evolution.

5.2 Explosive H-Burning

The major destruction of hydrogen in hydrostatic burning occurs via the pp-chain(s), initiated by $^1H(p, e^+\nu_e)^2H$, and the CNO cycle which converts ^1H into ^4He by a sequence of (p, γ) and (p, α) reactions on C, N, and O isotopes and subsequent beta-decays. Higher temperatures can enhance the reaction rates of charged-particle captures which depend on the penetration probability through Coulomb barriers. This can speed up the reactions following $^1H(p, e^+\nu_e)^2H$ and convert the pp-chains to the so-called hot pp-chains, which convert ^2H to CNO isotopes. However, this speeds up only the explosive burning of pre-existing ^2H, the destruction of the fuel ^1H is dependent on the weak reaction $^1H(p, e^+\nu_e)^2H$ which enforces long timescales. In the hydrostatic CNO-cycle the reaction $^{14}N(p, \gamma)^{15}O$ is the slowest reaction. Higher temperatures can speed up this fusion reaction (and the reaction $^{13}N(p, \gamma)^{14}O$ which then wins over the ^{13}N beta-decay) and lead to the hot CNO-cycle $^{12}C(p, \gamma)^{13}N(p, \gamma)^{14}O(e^+\nu_e)^{14}N(p, \gamma)^{15}O(e^+\nu_e)$. But this cycle is also limited by the beta-decay half-lives of ^{14}O and ^{15}O (of the

order of minutes) and does not permit H-destruction on the timescale of
seconds. The latter can only occur when nuclei above Ne are produced and
a sequence of proton captures and (very fast) beta-decays close to the proton drip-line converts hydrogen to heavy nuclei in the so-called rp-process
(rapid proton capture process). Such a process can be ignited in high
density environments, where the pressure is dominated by the degenerate
electron gas and shows no temperature dependence. This prevents a stable and controlled burning, leading therefore to a thermonuclear runaway[1].
While the ignition is always based on pp-reactions (as in solar hydrogen
burning), the runaway leads to the hot CNO-cycle, branching out partially
via $^{15}N(p,\gamma)^{16}O(p,\gamma)^{17}F(e^+\nu_e)^{17}O(p,\gamma)^{18}F$. In essentially all nuclei lighter
than Ca, a proton capture reaction on the nucleus $(Z_{even}-1, N = Z_{even})$ pro-
duces the compound nucleus above the alpha particle threshold and permits
a (p,α) reaction. This is typically not the case for $(Z_{even} - 1, N = Z_{even} - 1)$
due to the smaller proton separation energy and leads to hot CNO-type
cycles above Ne. There is one exception, $Z_{even} = 10$, where the reaction
$^{18}F(p,\alpha)$ is possible, avoiding ^{19}F and a possible leak via $^{19}F(p,\gamma)$ into the
NeNaMg-cycle. This has the effect that only alpha induced reactions like
$^{15}O(\alpha,\gamma)$ can aid a break-out from the hot CNO-cycle to heavier nuclei
beyond Ne. Break-out reactions from the hot CNO-type cycles above Ne
proceed typically via proton captures on the nucleus $(Z_{even}, N = Z_{even} - 1)$
and permits a faster build-up of heavier nuclei[2]. They occur at temperatures of about 3×10^8 K, while the alpha-induced break-out from the hot
CNO-cycle itself is delayed to about 4×10^8 K. Maximum temperatures
of the range 3×10^8 K provide major nucleosynthesis yields of hot CNO-
products like ^{15}N and specific nuclei between Ne and Si/S, which are based
on the processing of pre-existing Ne as discussed in caption of Figure 5.4.
Once the hot CNO-cycle, and higher cycles beyond Ne, have generated suf-

---

[1]Runaway thermonuclear reactions can occur in stars when nuclear fusion is ignited in
conditions under which the pressure exerted by overlying layers of the star greatly exceeds
thermal pressure, a situation that makes possible rapid increases in temperature. Such
a scenario may arise in stars containing degenerate matter, in which electron degeneracy
pressure rather than normal thermal pressure does most of the work of supporting the
star against gravity, and in stars undergoing implosion. In all cases, the imbalance arises
prior to fusion ignition; otherwise, the fusion reactions would be naturally regulated to
counteract temperature changes and stabilize the star. When thermal pressure is in equilibrium with overlying pressure, a star will respond to the increase in temperature and
thermal pressure due to initiation of a new exothermic reaction by expanding and cooling.
A runaway reaction is only possible when this response is inhibited.

[2]see F.-K. Thielemann, K.-L. Kratz, B. Pfeiffer, T. Rauscher, L. van Wormer, M. C.
Wiescher: Nucl. Phys. A 570, 329c (1994)

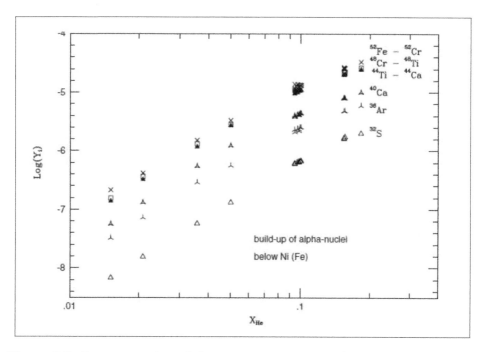

Figure 5.3: Demonstration of the composition up to Cr and from Mn to Ni (after decay) as a function of remaining alpha mass-fraction X_{He} from an alpha-rich freeze-out. Lighter nuclei, being produced by alpha-captures from a remaining alpha reservoir, have larger abundances for more pronounced alpha-rich freeze-outs.

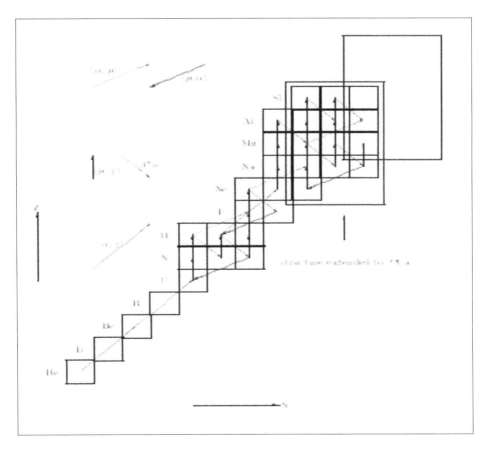

Figure 5.4: The hot CNO-cycle incorporates three proton captures,$(^{12}C, ^{13}N, ^{14}N)$, two β^+-decays $(^{14,15}O)$ and one (p, α)-reaction (^{15}N). In a steady flow of reactions the long beta-decay half-lives are responsible for high abundances of $^{15,14}N$ (from $^{15,14}O$ decay) in nova ejecta. From Ne to Ca, cycles similar to the hot CNO exist, based always on alpha-nuclei like ^{20}Ne, ^{24}Mg etc. The exception is the (not completed) cycle based on ^{16}O, due to $^{18}F(p, \alpha)^{15}O$, which provides a reaction path back into the hot CNO-cycle. Thus, in order to proceed from C to heavier nuclei, alpha-induced CNO break-outs are required. The shown flow pattern, which includes alpha-induced reactions, applies for temperatures in the range $4 - 8 \times 10^8 K$. Smaller temperatures permit already processing of pre-existing Ne via hot CNO-type cycles. This leads to the typical nova abundance pattern with $^{15,14}N$ enhancements, combined with specific isotopes up to about Si or even Ca. If the white dwarf is an ONeMg rather than CO white dwarf, containing Ne and Mg in its initial composition resulting from carbon rather than helium burning, the latter feature is specifically recognizable.

ficient amounts of energy in order to surpass temperatures of $\sim 4 \times 10^8$ K, alpha-induced reactions lead to a break-out from the hot CNO-cycle. This provides new fuel for reactions beyond Ne, leading to a further increase in temperature and subsequently also helium burns explosively in a thermonuclear runaway [3]. In the next stage of the ignition process also He is burnt the 3α-reaction $^4He(2\alpha,\gamma)^{12}C$, filling the CNO reservoir, and the αp-process (a sequence of (α, p) and (p, γ) reactions) sets in. It produces nuclei up to Ca and provides seed nuclei for hydrogen burning via the rp-process (proton captures and beta-decays). Processing of the α p-process and rp-process up to and beyond ^{56}Ni is depicted in Figure 5.5. Certain nuclei play the role of long waiting points in the reaction flux, where long beta-decay half-lives dominate the flow, either competing with slow (α, p) reactions or negligible (p, γ) reactions, because they are inhibited by inverse photodisintegrations for the given temperatures. Such nuclei were identified as $^{25}Si(\tau_{1/2} = 0.22s),^{29} S(0.187s),^{34} Ar(0.844s),^{38} Ca(0.439s)$. The bottle neck at ^{56}Ni can only be bridged for minimum temperatures around 10^9K (in order to overcome the Coulomb barrier for proton capture) and maximum temperatures below 2×10^9K (in order to avoid photo disintegrations), combined with high densities exceeding $10^6 gcm^{-3}$ which support the capture process. If this bottle neck can be overcome, other waiting points like $^{64}Ge(64s),^{68} Se(96s),^{74} Kr(17s)$ seem to be hard to pass. However, partially temperature dependent half-lives (due to excited state population), or mostly 2p-capture reactions via an intermediate proton-unstable nucleus can support. The final endpoint of the rp-process was found to be caused by a closed reaction cycle in the Sn-Sb-Te region due to increasing alpha-instability of heavy proton-rich nuclei[4].

5.3 Explosive He-Burning

Explosive He-burning is characterized by the same reactions as hydrostatic He-burning, producing ^{12}C and ^{16}O. Figure 5.1 indicates that even for temperatures beyond 10^9K high densities $(> 10^5 gcm^{-3})$ are required to burn essential amounts of He. This would cause major fuel destruction only in (electron) degenerate high density environments like white dwarfs with unburned He. Such environments are envisioned in sub-Chandrasekhar mass type Ia

[3]H. Schatz, A. Aprahamian, J. Görres, M. Wiescher, T. Rauscher, J.F. Rembges, F.-K. Thielemann, B. Pfeiffer, P. Möller, K.-L. Kratz, H. Herndl, B.A. Brown, H. Rebel: Phys. Rep. 294, 168 (1998)

[4]H. Schatz, A. Aprahamian, V. Barnard, L. Bildsten, A. Cumming, M. Ouel- lette, T. Rauscher, F.-K. Thielemann, M. Wiescher: Phys. Rev. Lett. 86, 3471 (2001)

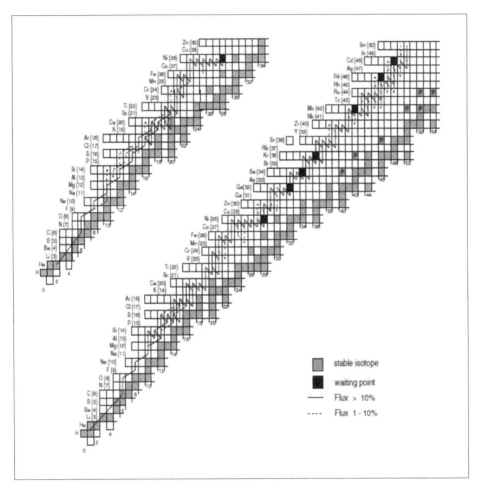

Figure 5.5: Rp and αp-process flows up to and beyond Ni. The reaction flows shown in the nuclear chart are integrated reaction fluxes from a time dependent network calculation (M. Wiescher, H. Schatz, A. Champagne: Phil. Trans. Roy. Soc. 356, 2105 (1998)), (a) during the initial burst and thermal runaway phase of about 10s (top left), (b) after the onset of the cooling phase when the proton capture on ^{56}Ni is not blocked anymore by photo disintegrations (extending for about 200s, bottom right). Waiting points above ^{56}Ni are represented by filled squares, stable nuclei by hatched squares, light p-process nuclei below A=100 are indicated by a P.

supernovae, but observations do not favor such models. During the passage of a $10^{51}erg$ (type II) supernova shock front through the He-burning zones of a massive $25M_\odot$ star, maximum temperatures of only $(6-9) \times 10^8$K are attained and the amount of He burnt is negligible. However, neutron sources like $^{22}Ne(\alpha,n)^{25}$Mg [or $^{13}C(\alpha,n)^{16}$O], which sustain an s-process neutron flux in hydrostatic burning, release a large neutron flux under explosive conditions. This leads to partial destruction of ^{22}Ne and the buildup of ^{25,26}Mg via $^{22}Ne(\alpha,n)^{25}Mg(n,\gamma)^{26}Mg$. Similarly, ^{18}O and ^{13}C are destroyed by alpha-induced reactions. This releases neutrons with $Y_n \simeq 2 \times 10^{-9}$ at a density of $8.3 \times 10^3 gcm^{-3}$, corresponding to $n_n \simeq 10^{19}cm^{-3}$ for about 0.2s, and causes neutron processing.

5.4   Explosive C- and Ne-Burning

The main burning products of explosive neon burning are ^{16}O, ^{24}Mg, and ^{28}Si, synthesized via the reactions $^{20}Ne(\gamma,\alpha)^{16}O$ or $^{20}Ne(\alpha,\gamma)^{24}Mg(\alpha,\gamma)^{28}Si$, similar to the hydrostatic case. The mass zones in supernovae which undergo explosive neon burning must have peak temperatures in excess of $2.1 \times 10^9 K$. They undergo a combined version of explosive neon and carbon burning (see Figures 5.1 and 5.2). Mass zones which experience temperatures in excess of $1.9 \times 10^9 K$ will undergo explosive carbon burning, as long as carbon fuel is available. This is often not the case in type II supernovae originating from massive stars. Besides the major abundances, mentioned above, explosive neon burning supplies also substantial amounts of ^{27}Al, ^{29}Si, ^{32}S, ^{30}Si, and ^{31}P. Explosive carbon burning contributes in addition the nuclei ^{20}Ne, ^{23}Na, ^{24}Mg, ^{25}Mg, and ^{26}Mg. Many nuclei in the mass range 20<A<30 can be reproduced in solar proportions. This was confirmed for realistic stellar conditions by Morgan[5] [275,438,375,377,314] and some others. As photo disintegrations become important in explosive Ne-burning, also heavier pre-existing nuclei in such burning shells, from previous s- or r-processing (originating from prior stellar evolution or earlier stellar generations), can undergo e.g. (γ,n) or (γ,α) reactions. These can produce rare proton-rich stable isotopes of heavy elements in the so-called p- or gamma-process.

5.5   Explosive O-Burning

Temperatures in excess of roughly $3.3 \times 10^9 K$ lead to a quasi-equilibrium (QSE) in the lower QSE-cluster which extends over the range 28<A<45 in

---

[5]J.A. Morgan: Astrophys. J. 238, 674 (1980)

mass number, while the path to heavier nuclei is blocked by small Q-values and reaction cross sections for reactions out of closed shell nuclei with Z or N=20 [429,157,158]. A full NSE with dominant abundances in the Fe-group cannot be attained. The main burning products are $^{28}Si, ^{32}S, ^{36}Ar, ^{40}Ca, ^{38}Ar$, and ^{34}S, while $^{33}S, ^{39}K, ^{35}Cl, ^{42}Ca$, and ^{37}Ar have mass fractions of less than 10^{-2}. In zones with temperatures close to 4×10^9K there exists some contamination by the Fe-group nuclei $^{54}Fe, ^{56}Ni, ^{52}Fe, ^{58}Ni, ^{55}Co$, and ^{57}Ni. The abundance distribution within the QSE-cluster is determined by alpha, neutron, and proton abundances. Because electron captures during explosive processing are negligible, the original neutron excess stays unaltered and fixes the neutron to proton ratio. Under those conditions the resulting composition is dependent only on the alpha to neutron ratio at freeze-out. In an extensive study Woosey et al[6] noted that with a neutron excess $\eta (= \sum_i (N_i - Z_i)Y_i)$ of 2×10^{-3} the solar ratios of $^{39}K/^{35}Cl, ^{40}Ca/^{36}Ar, ^{36}Ar/^{32}S, ^{37}Cl/^{35}Cl, ^{38}Ar/^{34}S, ^{42}Ca/^{38}Ar, ^{41}K/^{39}K$, and $^{37}Cl/^{33}S$ are attained within a factor of 2 for freeze-out temperatures in the range $(3.1 - 3.9) \times 10^9 K$. This is the typical neutron excess resulting from solar CNO-abundances, which are first transformed into ^{14}N in H-burning and then into ^{22}Ne in He-burning via $^{14}N(\alpha, \gamma)^{18}F(\beta^+)^{18}O(\alpha, \gamma)^{22}Ne$, while for lower values of the neutron excess (as expected for stars of lower metallicity) essentially only the alpha nuclei $^{28}Si, ^{32}S, ^{36}Ar, ^{40}Ca$ are produced in sufficient amounts. This affects elemental abundances and causes an odd-even staggering in Z. There exist extensive discussions of such results in supernova models in the literature.

5.6 Explosive Si-Burning

Zones which experience temperatures in excess of $4.0 - 5.0 \times 10^9 K$ undergo explosive Si-burning. For $T >> 5 \times 10^9 K$ essentially all Coulomb barriers can be overcome and a nuclear statistical equilibrium is established. Such temperatures lead to complete Si-exhaustion and produce Fe-group nuclei. The doubly-magic nucleus ^{56}Ni, with the largest binding energy per nucleon for N=Z, is formed with a dominant abundance in the Fe-group in case Y_e is larger than 0.49. Explosive Si-burning can be divided into three different regimes: (i) incomplete Si-burning and complete Si-burning with either (ii) a normal or (iii) an alpha-rich freeze-out. Which of the three regimes is encountered depends on the peak temperatures and densities attained during the passage of a supernova shock/burning front shown in Figure 5.6). Researchers recognizes that the mass zones of SNe Ia and SNe II expe-

---

[6]S.E. Woosley, W.D. Arnett, D.D. Clayton: Astrophys. J. Suppl. 26, 231 (1973)

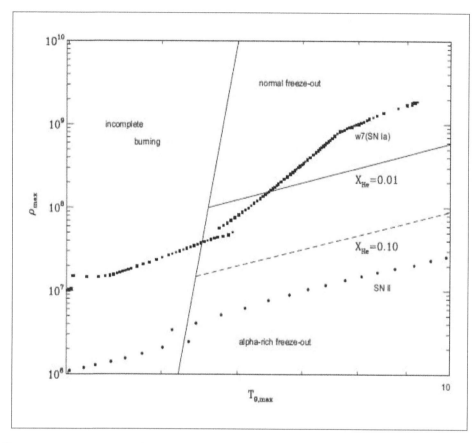

Figure 5.6: Division of the $\rho_{max} - T_{max}$ -plane for adiabatic expansions from ρ_{max} and T_{max} with an adiabatic index of 4/3 and a hydrodynamic timescale equal to the free fall timescale. Conditions separate into incomplete and complete Si-burning with normal and alpha-rich freeze-out. Contour lines of constant ^{4}He mass fractions in complete burning are given for levels of 1 and 10%. They coincide with lines of constant radiation entropy per gram of matter. For comparison also the maximum ρ - T-conditions of individual mass zones in type Ia and type II supernovae are indicated.

rience different regions of complete Si-burning. At high temperatures in complete Si-burning or also during a normal freeze-out, the abundances are in a full nuclear statistical equilibrium (NSE) and given by Eq. (4.18). An alpha-rich freeze-out is caused by the inability of the triple-alpha reaction $^4He(2\alpha,\gamma)^{12}C$, transforming ^4He into ^{12}C, and the $^4He(\alpha n,\gamma)^9Be$ reaction, to keep light nuclei like n, p, and ^4He, and intermediate mass nuclei beyond A=12 in an NSE during declining temperatures, when the densities are small. The latter enter quadratically for these rates, causing during the fast expansion and cooling in explosive events a large alpha abundance after charged particle freeze-out, which shifts the QSE groups to heavier nuclei, transforming e.g. $^{56}Ni,^{57}Ni$, and ^{58}Ni into $^{60}Zn,^{61}Zn$, and ^{62}Zn. This also leads to a slow supply of carbon nuclei still during freeze-out, leaving traces of alpha nuclei, $^{32}S,^{36}Ar,^{40}Ca,^{44}Ti,^{48}Cr$, and ^{52}Fe, which did not fully make their way up to ^{56}Ni. Figure 5.3 shows this effect, typically for SNe II, as a function of remaining alpha-particle mass fraction after freeze-out. The major NSE nuclei $^{56}Ni,^{57}Ni$, and ^{58}Ni get depleted when the remaining alpha fraction increases, while all other species mentioned above increase. Incomplete Si-burning is characterized by peak temperatures of $(4-5) \times 10^9 K$. Temperatures are not high enough for an efficient bridging of the bottle neck above the proton magic number Z = 20 by nuclear reactions. Besides the dominant fuel nuclei ^{28}Si and ^{32}S we find the alpha-nuclei ^{36}Ar and ^{40}Ca being most abundant. Partial leakage through the bottle neck above Z=20 produces ^{56}Ni and ^{54}Fe as dominant abundances in the Fe-group. Smaller amounts of $^{52}Fe,^{58}Ni,^{55}Co$, and ^{57}Ni are encountered. The above discussion focuses mostly on conditions for $Y_e \simeq 0.5$. The results can vary strongly with more proton or more neutron rich environments, which also has an influence on Figures 5.3 and 5.6. Explosive Si-burning occurs in type Ia as well as in type II supernovae, and both environments encounter physical effects which can alter Y_e drastically (electron captures at high densities and temperatures in SNe Ia as well as II, neutrino-induced reactions on nucleons and nuclei in SNe II). The basic recipe for SN Ia explosions is simple. A type Ia supernova results from runaway carbon fusion in the core of a carbon-oxygen white dwarf star. If a white dwarf, which is composed almost entirely of degenerate matter (e.g., electron gas), can gain mass from a companion, the increasing temperature and density of material in its core will ignite carbon fusion if the star's mass approaches the Chandrasekhar limit. This leads to an explosion that completely disrupts the star due to thermonuclear runaway. Luminosity increases by a factor of greater than 5 billion. One way to gain the additional mass would be by accreting gas from a giant star (or even main sequence) companion. A second and apparently more common mechanism to generate the same type of explosion is the merger of two white dwarfs. All explosive burning phases outlined above will be discussed in more detail for a number of astrophysical events in Chapters 6 through 8.

5.7 The r-Process

The astrophysical r-process is a result of explosive Si-burning for either extreme conditions in Y_e and/or the entropy of the matter involved. The operation of an r-process is characterized by the fact that 10 to 100 neutrons per seed nucleus (in the Fe-peak or somewhat beyond) have to be available to form all heavier r-process nuclei by neutron capture. For a composition of Fe group nuclei and free neutrons that translates into a neutron excess of $\eta = \sum_i (N_i - Z_i)Y_i = 0.4 - 0.7$ or $Y_e = \sum_i Z_i Y_i = 0.15 - 0.3$. Such a high neutron excess can only be available if weak interactions play a major role either via electron capture or neutrino absorption on nucleons and nuclei. In the case of electron capture one needs energetic electrons (on protons or nuclei) which have to overcome large negative Q-values. This can be achieved by degenerate electrons with large Fermi energies and requires a compression to densities of $10^{11} - 10^{12} gcm^{-3}$, with a β equilibrium between electron captures and β^--decays [68] as found in neutron star matter. Another option is an extremely alpha-rich freeze-out in complete Si-burning with moderate neutron excesses η and Y_e's. After the freeze-out of charged particle reactions in matter which expands from high temperatures but relatively low densities, 70 to 95% of all matter can be locked into ^4He with N=Z. Figure 5.6 exhibited the onset of such an extremely alpha rich freeze-out by indicating contour lines for He mass fractions of 1 and 10%. These contour lines correspond to T_9^3/ρ=const, which is proportional to the entropy per gram of matter of a radiation dominated gas. Thus, the radiation entropy per gram of baryons can be used as a measure of the ratio between the remaining He mass-fraction and heavy nuclei. Similarly, the ratio of neutrons to Fe-group (or heavier) nuclei (i.e. the neutron to seed ratio) is a function of entropy and permits for high entropies, with large remaining He and neutron abundances and small heavy seed abundances, neutron captures which proceed to form the heaviest r-process nuclei). A different situation arises for maximum temperatures below freeze-out conditions for charged particle reactions with Fe-group nuclei. In that situation reactions among light nuclei which release neutrons, like (α, n) reactions on ^{13}C and ^{22}Ne, can sustain a neutron flux. The constraint of having 10-100 neutrons per heavy nucleus, in order to attain r-process conditions, can only be met by small abundances of Fe-group nuclei. Such conditions were expected when a shock front passes the He-burning shell and enhances the $^{22}Ne(\alpha, n)$ reaction by orders of magnitude. However, it was found that this neutron source is not strong enough for an r-process in realistic stellar models. Researches, based on additional neutron release via inelastic neutrino scattering also could not produce neutron densities which are required for such a process to become operative. For a number of years r-process calculations, independent of a specific astrophysical site, and just based on the goal to find the required neutron number densities and temperatures which can reproduce the solar abundance pattern of heavy elements, have been performed, which together with applications to the astrophysical sites listed above, will be discussed in Chapter 7.

Exercise

1. The inverse β-decay reaction $\nu_e +^{37}Cl \rightarrow e^- +^{37}Ar$ is used to detect neutrinos from the sun. The number of solar neutrinos produced may be estimated from the solar constant ($1350W/m^2$). Assume that 10% of the thermonuclear energy is carried away by neutrinos with a mean energy of 1 MeV each and that only about 1% of the neutrino is energetic enough to convert ^{37}Cl to ^{37}Ar. For a detector containing $400g/m^3$ of tetrachloroethylene (C_2Cl_4), estimate the average number of ^{37}Ar produced in a day if the density of C_2Cl_4 is $1.5g/cm^3$ and about a quarter of the chlorine is ^{37}Cl. The cross section for the reaction may be taken to be $10^{-48}m^2$.

2. Assuming a cross section of $10m^2$ for a neutrino to interact with each nucleon, find the difference in the night and daytime detection rate of solar neutrinos for the Super Kamiokande water Cerenkov detector consisting of 5×10^4 tons of water.

References

Quoted in the text:

- M. Wiescher, J. Görres, S. Graaf, L. Buchmann, F.-K. Thielemann: Astrophys.J. 343, 352 (1989)

- F.-K. Thielemann, K.-L. Kratz, B. Pfeiffer, T. Rauscher, L. vanWormer, M. C.Wiescher: Nucl. Phys. A 570, 329c (1994)

- M. Wiescher, J. Görres, H. Schatz, J. Phys. G: Nucl. Part. Phys. 25, R133 (1999)

- F. Rembges, C. Freiburghaus, T. Rauscher, F.-K. Thielemann, H. Schatz, M. Wiescher; Astrophys. J. 484, 412 (1997)

- E. Livne, W.D. Arnett: Astrophys. J. 452, 62 (1995)

- S. Goriely, J. Jose, M. Hernanz, M. Rayet, M. Arnould: Astrophys. and Astron.383, 27 (2002)

- F.-K. Thielemann, M. Arnould, W. Hillebrandt: Astron. and Astrophys. 74,175 (1979)

- F.-K. Thielemann, M. Arnould, J. W.Truran: In: Advances in Nuclear Astrophysics, ed. E. Vangioni-Flam (Editions Frontiere, Gif sur Yvette 1987) p. 525

- F.-K. Thielemann, F. Brachwitz, P. Höflich, G. Martinez-Pinedo, K. Nomoto: New Astronomy 48, 65 (2004)

- F.-K. Thielemann, J. Metzinger, H.V. Klapdor: Z. Phys. A 309, 301 (1983)

- J.W. Truran, J.J. Cowan, A.G.W. Cameron: Astrophys. J. 222, L63 (1978)

- S.E. Woosley, W.M. Howard: Astrophys. J. Suppl. 36, 285 (1978)

Chapter 6

Core Collapse Supernovae

6.1 Introduction

Supernovae are generally produced in two different explosion mechanisms. The one via the thermonuclear explosion of a white dwarf which has been accreting matter from a companion is known as a Type Ia supernova. While in the second instance , core-collapse of massive stars produce Type II, Type Ib and Type Ic supernovae. All stars, irrespective of their mass, progress through the first stages of their lives in a similar way, by converting hydrogen into helium. As the hydrogen is used up, fusion reactions slow down resulting in the release of less energy, and gravity forces the core to contract. This contraction raises the temperature of the core once again, generally to the point where helium fusion can begin. Once helium has been used up, the core contracts again, and in low-mass stars this is where the fusion processes end – with the creation of an electron degenerate carbon core. For massive ($>10M_\odot$) stars, however, this is not the end. The contraction of the helium core raises the temperature sufficiently so that carbon burning can begin. After the carbon burning stage comes the neon burning, oxygen burning and silicon burning stages, each lasting a shorter period of time than the previous one as depicted in Figure 6.1. The end result of the silicon burning stage is the production of iron, and it is this process which spells the end for the star. Just before core-collapse, the interior of a massive star resembles the layer structure of an onion, with shells of successively lighter elements burning around an iron core. These burning stages become shorter and shorter as lighter elements are fused into heavier elements. Up until this stage, the enormous mass of the star has been supported against gravity by the energy released in fusing lighter elements into heavier ones. Iron, however, is

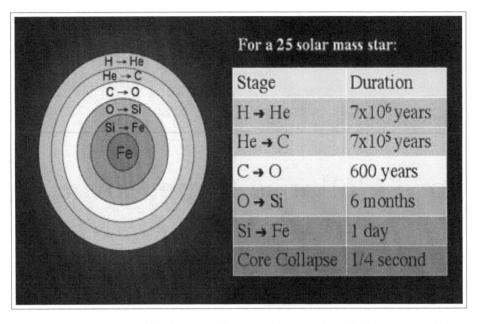

Figure 6.1: The onion-like layers of a massive star just before core collapse. [https://en.wikipedia.org/wiki/*Type_II_supernova*]

the most stable element and must actually absorb energy in order to fuse into heavier elements. The formation of iron in the core therefore effectively concludes fusion processes and, with no energy to support it against gravity, the star begins to collapse in on itself. The star has less than 1 second of life remaining. During this final second, the collapse causes temperatures in the core to skyrocket, which releases very high-energy gamma rays. These photons undo hundreds of thousands of years of nuclear fusion by breaking the iron nuclei up into helium nuclei in a process called photo disintegration. At this stage the core has already contracted beyond the point of electron degeneracy, and as it continues contracting, protons and electrons are forced to combine to form neutrons. This process releases vast quantities of neutrinos carrying substantial amounts of energy, again causing the core to cool and contract even further. The contraction is finally halted once the density of the core exceeds the density at which neutrons and protons are packed together inside atomic nuclei. It is extremely difficult to compress matter beyond this point of nuclear density as the strong nuclear force becomes repulsive. Therefore, as the innermost parts of the collapsing core overshoot this mark, they become slow in their contraction and ultimately rebound. This creates an outgoing shock wave which reverses the in falling motion of the material in the star and accelerates it outwards. Aiding in the propagation of this shock wave through the star are the neutrinos which are being created in massive quantities under the extreme conditions in the core. Under normal circumstances neutrinos interact very weakly with matter, but under the extreme densities of the collapsing core, a small fraction of them can become trapped behind the expanding shock wave. The energy of these trapped neutrinos increases the temperature and pressure behind the shock wave, which in turn gives it strength as it moves out through the star. The passage of this shock wave compresses the material in the star to such a degree that a whole new wave of nucleosynthesis occurs. These reactions produce many more elements including all the elements heavier than iron, a feat the star was unable to achieve during its lifetime. The creation of such elements requires an enormous input of energy and core-collapse supernovae are one of the very few places in the Universe where such energy is available. Eventually, after a few hours, the shock wave reaches the surface of the star and expels stellar material and newly created elements into the interstellar medium. What is left behind is either a neutron star or a black hole depending on the final mass of the core. Core collapse supernovae are the spectacular explosions (see Figure 6.2) that mark the violent death of massive stars. These events are the most energetic explosions in the cosmos, releasing energy of order 10^{53} erg at the staggering rate of 10^{45-46}. The $^{12}C(\alpha, \gamma)^{16}O$ reaction causes large uncertainties in stellar evolution and for the pre-supernova models[1]. The treatment of convection in stellar evolution , especially the issue of overshooting and semi-convection is has yet not settled. This influence the possible growth of the He-burning core, which is responsible for the mixing in of fresh He at higher temperature and enhancement of the O/C ratio. Stellar evolution calculations indicated that the total amount of ^{16}O can also vary by almost a factor of 3, for the extreme choices of the semi-convection parameter. The resulting properties in terms of size and composition of the stellar core after He-burning have a strong influence on all subsequent stages of burning. At the end of hydrostatic burning, a massive star consists of concentric shells that are the remnants of its previous burning phases (i.e, hydrogen, helium, carbon, neon, oxygen, silicon burning phases). Iron is the final stage of nuclear fusion in hydrostatic burning, as the synthesis of any heavier element from lighter elements does not release energy beyond formation of iron; rather, energy must be used up. When the iron core, formed in the center of the massive star, exceeds the Chandrasekhar mass limit of about 1.44 solar masses , electron degeneracy

---

[1]B.W. Filippone, J. Humblet, K. Langanke: Phys. Rev. C 40, 515 (1989)

Figure 6.2: Cassiopeia A -The colorful aftermath of a violent stellar death.[Credit: http://www.whillyard.com]

pressure can no longer stabilize the core and it collapses, starting what is called a core collapse supernova and in most of the cases if the hydrogen envelope was not lost during stellar evolution this coincides with a type II supernova, which in most cases results from the collapse of the core of large stars at least 8 or 9 solar masses. Type II supernovae occur in three broad types- Type-IIb, type -IIL and type-IIP.

- Type IIL: The type of supernova has a light curve that decreases over time in a linear fashion after reaching maximum luminosity. SN 1979C, in the galaxy M100, is an example that is interesting as, though it has faded predictably in visible light, it has maintained its output in X-Rays. It is the brightest X-Ray source in its galaxy, outside the galaxy center. The progenitor star had a mass of about 18 solar masses, and had a very strong stellar wind. This image shows the supernova, which is labeled, and the containing galaxy M100. It is a composite image with the infrared image from Spitzer in red, Chandra's X-ray image in gold, and optical imagery in red, green, and blue.

- Type IIb: This is very similar to a Type Ib supernova except that some hydrogen remains in the outer layers of the star, and is seen in the star's spectrum, which is not the case for the type Ib. Cassiopeia A is the remnant of a type IIb supernova that occurred approximately 11,000 light-years away. The expanding debris cloud of material left over from the supernova is now approximately 10 light-years across, and is expanding at between 4,000 and 6,000 km/second.

- Type IIP: The light curve for the type IIP decays more slowly, then flattens for a while creating a plateau region before continuing to decrease. An example is SN 1987A, though some observers suggest this was a quark nova. The first image shows it in the Tarantula Nebula in the Large Magellanic Cloud. The second image shows a series of pictures of the ejecta from the explosion over a period of about twelve years. Interestingly, as at January 2015, no visible evidence of a neutron star remnant has been found.

The core collapse, forces protons and electrons together to produce neutrons and electron neutrinos. The neutrinos escape, as they interact so feebly with other matter, and take energy away from the core, which accelerates the collapse. During the collapse, the outer parts of the core are moving inwards at up to 23% of the speed of light, so the process takes only a few milliseconds. When the collapse is halted by neutron degeneracy pressure, the in falling layers bounce off the core, throwing off all the outer material, producing the supernova. At this stage, the temperature at the center of the core is about 100 billion degrees kelvin, so thermal neutrinos are produced as neutrino/anti-neutrino pairs of all three flavors; electron, muon and tau neutrinos. This neutrino burst lasts about 10 seconds, and produces a stupendous blast of energy. The time from when the star starts to burn its helium until its death is about 10% of the time the star spends on the main sequence, burning hydrogen. If the star started with more than about $20M_\odot$ solar masses, it is likely to continue to collapse, overcoming the neutron degeneracy pressure, to form a black hole. This will accrete some of the matter that bounced of the neutron core over a period of time. If the star started with more than about 40 or 50 solar masses, the "bounce" off the neutron core is skipped, avoiding the supernova stage, and the star collapses immediately into a black hole. In its aftermath a central neutron star or black hole is formed and the surrounding shells are ejected into the Interstellar Medium. Although this general picture has been confirmed by the various observations from supernova SN1987A, simulations of the core collapse and the explosion are still far from being completely understood and robustly modeled. To enrich the input which goes into the simulation of type II supernovae and to improve the models and their numerical simulations is a very active research field at various institutions worldwide. The collapse is very sensitive to the entropy and to the number of leptons per baryon, Y_e [2]. In turn these two quantities are mainly determined by the weak β-decay and electron-capture processes. First, in the early stage of the collapse Y_e is reduced as it is energetically favorable to capture electrons, which at the densities involved have Fermi energies of a few MeV, on (Fe-peak) nuclei. This reduces the electron pressure, thus accelerating the collapse, and shifts the distribution of nuclei present in the core to more neutron-rich material. Second, many of the nuclei present can also β decay. While this process is quite unimportant compared to electron capture for initial Ye values around 0.5, it becomes increasingly competitive for neutron rich nuclei due to an increase in phase space related to larger Q_β values.

6.2   General Scenario

The weak processes like *beta*-decay, electron capture, and photo disintegration cost the core energy and reduce its electron density. As a consequence, the collapse is accelerated. An important change in the physics of the collapse occurs, as the density reaches $\rho_{trap} \simeq 4.0 \times 10^{11} gcm^{-3}$. Then neutrinos are essentially trapped in the core, as their diffusion time (due to coherent elastic scattering on nuclei) becomes larger than the collapse time. After neutrino trapping, the collapse proceeds homologous, until nuclear densities ($\rho N \simeq 10^{14} gcm^{-3}$) are reached. As nuclear matter has a finite compressibility, the homologous core decelerates and bounces in response to the increased nuclear matter pressure; this eventually drives an outgoing shock wave into the outer core; i.e. the envelope of the iron core outside the homologous core, which in the meantime has continued to fall inwards at supersonic speed. The core bounce with the formation of a shock wave is believed to be the mechanism that triggers a supernova explosion, but several ingredients of this physically appealing picture and the actual mechanism of a supernova explosion are still uncertain and controversial. If the shock wave is strong enough not only to stop the collapse, but also to explode the outer burning shells of the star, one speaks about the 'prompt mechanism'[3]. However, it appears as if the energy available to the shock is not sufficient, and the shock will store its energy in the outer core, for example, by dissociation of nuclei into nucleons. Furthermore, this change

---

[2]H.A. Bethe, G.E. Brown, J. Applegate, J.M. Lattimer: Nucl. Phys. A 324, 487 (1979)
[3]J.R. Wilson: In Numerical Astrophysics, ed. by J.M. Centrella, J.M. LeBlanc, R.L. Bowers, (Jones and Bartlett, Boston 1985) p. 422

in composition results to additional energy losses, as the electron capture rate on free protons is significantly larger than on neutron-rich nuclei due to the smaller Q-values involved. This leads to a further neutronization of the matter. Part of the neutrinos produced by the capture on the free protons behind the shock leave the star carrying away energy. After the core bounce, a compact remnant is left behind. Depending on the stellar mass, this is either a neutron star (masses roughly smaller than 30 solar masses) or a black hole. The neutron star remnant is very lepton-rich (electrons and neutrinos), the latter being trapped as their mean free paths in the dense matter is significantly shorter than the radius of the neutron star. It takes a fraction of a second for the trapped neutrinos to diffuse out, giving most of their energy to the neutron star during that process and heating it up. The cooling of the protoneutron star then proceeds by pair production of neutrinos of all three generations which diffuse out. After several tens of seconds the star becomes transparent to neutrinos and the neutrino luminosity drops significantly. In the 'delayed mechanism', the shock wave can be revived by the outward diffusing neutrinos, which carry most of the energy set free in the gravitational collapse of the core and deposit some of this energy in the layers between the nascent neutron star and the stalled prompt shock. This lasts for a few 100 ms, and requires about 1% of the neutrino energy to be converted into nuclear kinetic energy. The energy deposition increases the pressure being the shock and the respective layers begin to expand, leaving between shock front and neutron star surface a region of low density, but rather high temperature. This region is called the hot neutrino bubble. The persistent energy input by neutrinos keeps the pressure high in this region and drives the shock outwards again, eventually leading to a supernova explosion. It has been found that the delayed supernova mechanism is quite sensitive to physics details deciding about success or failure in the simulation of the explosion. Two quite distinct improvements have been proposed viz. convective energy transport and in-medium modifications of the neutrino opacities which increase the efficiency of the energy transport to the stalled shock. Current one-dimensional supernova simulations, including sophisticated neutrino transport, fail to explode. The interesting question is whether the simulations explicitly requires multidimensional effects like rotation, magnetic fields or convection or the micro physics input in the one-dimensional models is insufficient. It is the goal of nuclear astrophysics to improve on this micro physics input. The next section shows that core collapse supernovae are nice examples to demonstrate how important microphysics input and progress in nuclear modeling can be. Late-stage stellar evolution is described in two steps. In the presupernova models the evolution is studied through the various hydrostatic core and shell burning phases until the central core density reaches values up to $10^{10} gcm^{-3}$. The models consider a large nuclear reaction network. However, the densities involved are small enough to treat neutrinos solely as an energy loss source. For even higher densities this is no longer true as neutrino-matter interactions become increasingly important. In modern core collapse codes neutrino transport is described self-consistently by spherically symmetric multi group Boltzmann simulations. While this is computationally very challenging, collapse models have the advantage that the matter composition can be derived from Nuclear Statistical Equilibrium (NSE) as the core temperature and density are high enough to keep reactions mediated by the strong and electromagnetic interactions in equilibrium. This means that for sufficiently low entropies, the matter composition is dominated by the nuclei with the highest Q-values for a given Y_e. The presupernova models are the input for the collapse simulations which follow the evolution through trapping, bounce and hopefully explosion.

6.3   Role of weak-Interaction rates in presupernova evolution

The collapse is the result of the tug of war between the two weakest known forces of nature- gravity and weak interaction. The weak processes like the capture of electrons by nuclei and protons and beta decay plays vital role during silicon burning. These processes become important when nuclei with masses $A \sim 55 - 60$ (pf-shell nuclei) are most abundant in the core (although capture on sd shell nuclei has to be considered as well). As weak interactions changes Y_e and electron capture dominates, the Y_e value is successively reduced from its initial value ~ 0.5. As a consequence, the abundant nuclei become more neutron rich and heavier, as nuclei with decreasing Z/A ratios are more bound in heavier nuclei. It is remarkable i) that there are many nuclei with appreciable abundances in the cores of massive stars during their final evolution, ii) neither the nucleus with the largest capture rate nor the most abundant one are necessarily the most relevant for the dynamical evolution, but it is the product of rate times abundance what makes a nucleus relevant. Figure 6.3 exhibits the consequences of the shell model weak interaction rates for presupernova models in terms of the three decisive quantities: the central Y_e value, entropy and the iron core mass. The central values of Y_e at the onset of core collapse increased by a significant rate of 0.01-0.015. It is important to note that the new models also result in

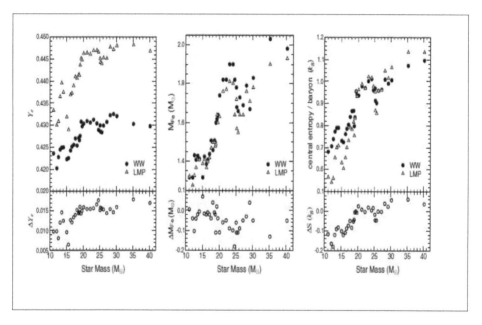

Figure 6.3: Comparison of the center values of Y_e (left), the iron core sizes (middle) and the central entropy (right) for $11 - 40 M_\odot$ stars between the WW models, which used the FFN rates, and the ones using the shell model weak interaction rates (LMP).

lower core entropies for stars with $M \leq 20 M_\odot$, while for $M \geq 20 M_\odot$, the new models actually have a slightly larger entropy. The iron core masses are generally smaller in the new models where the effect is larger for more massive stars ($M \geq 20 M_\odot$), while for the most common supernovae ($M \leq 20 M_\odot$) the reduction is by about $0.05 M_\odot$. Electron capture dominates the weak-interaction processes during presupernova evolution. However, during silicon burning, β decay (which increases Y_e) can compete and adds to the further cooling of the star. With increasing densities, β-decays are hindered as the increasing Fermi energy of the electrons blocks the available phase space for the decay. Thus, during collapse β-decays can be neglected. We note that the shell model weak interaction rates predict the presupernova evolution to proceed along a temperature-density-Y_e trajectory where the weak processes are dominated by nuclei rather close to stability. Thus it will be possible, after radioactive ion-beam facilities become operational, to further constrain the shell model calculations by measuring relevant beta decays and GT distributions for unstable nuclei. Heger et al[4] identified those nuclei which dominate the electron capture and beta decay during various stages of the final evolution of a $15 M_\odot, 25 M_\odot$ and $40 M_\odot$ star.

6.4   The Role of Electron Capture During Collapse

A very simple description for electron capture on nuclei has been used in collapse simulations estimating the rate in the light of independent particle model (IPM), considering pure Gamow-Teller (GT) transitions for only single particle states for proton and neutron numbers between N = 20 - 40[5]. More specifically this model assigns vanishing electron capture rates to nuclei with neutron numbers larger than N = 40, motivated by the observation that, within the IPM, GT_+ transitions are Pauli-blocked for nuclei with $N \geq 40$ and $Z \leq 40$[6]. However, as electron capture reduces Y_e, the nuclear composition is shifted to more neutron rich heavier nuclei, including those with N >40, which dominate the matter composition for densities larger than a few $10^{10} g cm^{-3}$. As a consequence of the model applied in the previous collapse simulations, electron capture on nuclei ceases at these densities and the capture is entirely due to free protons. This employed model for electron capture on nuclei is too simple and leads to incorrect

---

[4]A. Heger, K. Langanke, G. Martinez-Pinedo, S.E. Woosley: Phys. Rev. Lett. 86, 1678 (2001)

[5]S.W. Bruenn: Astrophys. J. Suppl. 58, 771 (1985); A. Mezzacappa, S.W. Bruenn: Astrophys. J. 405, 637 (1993); 410, 740 (1993

[6]G. M. Fuller: Astrophys. J. 252, 741 (1982)

conclusions, as the Pauli-blocking of the GT_+ transitions is overcome by correlations[7] and temperature effects. At first, the residual nuclear interaction, beyond the IPM, mixes the pf shell with the levels of the sdg shell, in particular with the lowest orbital, $g_{9/2}$. This makes the closed $g_{9/2}$ orbit a magic number in stable nuclei (N = 50) and introduces, for example, a very strong deformation in the N = Z = 40 nucleus ^{80}Zr. Moreover, the description of the $B(E2, 0^+ \rightarrow 2_1^+)$ transition in ^{68}Ni requires configurations where more than one neutron is promoted from the pf shell into the $g_{9/2}$ orbit, unblocking the GT_+ transition even in this proton magic N = 40 nucleus. Such a non-vanishing GT_+ strength has already been observed for ^{72}Ge (N = 40)[8] and ^{76}Se (N = 42)[9]. Secondly, during core collapse electron capture on the nuclei of interest here occurs at temperatures T > 0.8 MeV, which, in the Fermi gas model, corresponds to a nuclear excitation energy $U \simeq AT^2/8 > 5$ MeV; this energy is noticeably larger than the splitting of the pf and sdg orbitals ($E_{g_{9/2}} - E_{p_{1/2}}, f_{5/2} \simeq 3 MeV$). Hence, the configuration mixing of sdg and pf orbitals will be rather strong in those excited nuclear states of relevance for stellar electron capture. Furthermore, the nuclear state density at $E \sim 5 MeV$ is already larger than 100/MeV, making a state-by-state calculation of the rates impossible, but also emphasizing the need for a nuclear model which describes the correlation energy scale at the relevant temperatures appropriately. This model is the Shell Model Monte Carlo (SMMC) approach[10] which describes the nucleus by a canonical ensemble at finite temperature and employs a Hubbard-Stratonovich linearization of the imaginary-time many-body propagator to express observables as path integrals of one-body propagators in fluctuating auxiliary fields. Since Monte Carlo techniques avoid an explicit enumeration of the many-body states, they can be used in model spaces far larger than those accessible to conventional methods. The Monte Carlo results are in principle exact and are in practice subject only to controllable sampling and discretization errors. To calculate electron capture rates for nuclei A = 65–112 SMMC calculations have been performed in the full pf-sdg shell, using a residual pairing plus quadruple interaction, which, in this model space, reproduces well the collectivity around the N = Z = 40 region and the observed low-lying spectra in nuclei like ∗64Ni and ^{64}Ge. From the SMMC calculations the temperature-dependent occupation numbers of the various single-particle orbitals have been determined. These occupation numbers then became the input in RPA

[7]K. Langanke, E. Kolbe, D.J. Dean: Phys. Rev. C 63, 032801 (2001

[8]See M.C. Vetterli et al.: Phys. Rev. C 45, 997 (1992)

[9]R.L. Helmer et al.: Phys. Rev. C 55, 2802 (1997)

[10]S.E. Koonin, D.J. Dean, K. Langanke: Phys. Rep. 278, 1 (1997)

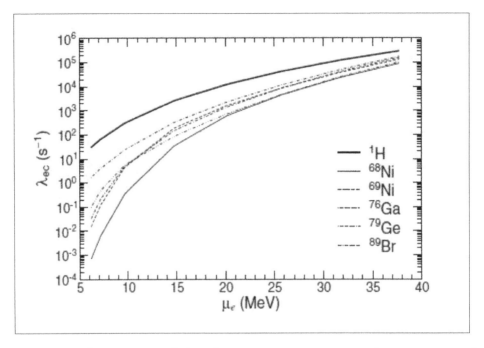

Figure 6.4: Comparison of the electron capture rates on free protons and selected nuclei as function of the electron chemical potential along a stellar collapse trajectory without considering neutrino blocking of the phase space in the calculation of the rates. Ref. Phys. Rev. Lett. 86, 1935 (2001)

calculations of the capture rate, considering allowed and forbidden transitions up to multi poles J = 4 and including the momentum dependence of the operators. The method has been validated against capture rates calculated from diagonalization shell model studies for $^{64,66}Ni$. For all studied nuclei one finds neutron holes in the (pf) shell and, for Z > 30, non-negligible proton occupation numbers for the sdg orbitals. This unblocks the GT transitions and leads to sizable electron capture rates. Figure 6.4 demonstrates the comparison of electron capture rates for free protons and selected nuclei along a core collapse trajectory. Depending on their proton-to-nucleon ratio Ye and their Q-values, these nuclei are abundant at different stages of the collapse. For all nuclei, the rates are dominated by GT transitions at low densities, while forbidden transitions contribute sizably at $\rho_{11} > 1$. Simulations of core collapse require reaction rates for electron capture on protons, $R_p = Y_p \lambda_p$, and nuclei $R_h = \sum_i Y_i \lambda_i$ (where the sum runs over all the nuclei present and Y_i denotes the number abundance of a given species), over wide ranges in density and temperature. Calculation of R_h requires knowledge of the nuclear composition, in addition to the electron capture rates described earlier. Saha-like NSE has been adopted to determine the needed abundances of individual isotopes and to calculate R_h and the associated emitted neutrino spectra on the basis of about 200 nuclei in the mass range A = 45-112 as a function of temperature, density and electron fraction. The rates for the inverse neutrino-absorption process are determined from the electron capture rates by detailed balance. Due to its much smaller $|Q|$-value, the electron capture rate on the free protons is larger than the rates of abundant nuclei during the core collapse (Figure 6.2). However, this is misleading as the low entropy keeps the protons significantly less abundant than heavy nuclei during the collapse. Figure 6.5 shows that the reaction rate on nuclei, R_h, dominates the one on protons, R_p, by roughly an order of magnitude throughout the collapse when the composition is considered. Only after the bounce shock has formed does R_p become higher than R_h, due to the high entropies and high temperatures in the shock-heated matter that result in a high proton abundance. The obvious conclusion is that electron capture on nuclei must be included in collapse simulations. It is also important to stress that electron capture on nuclei and on free protons differ quite noticeably in the neutrino spectra they generate. This is demonstrated in Figure 6.5 which shows that neutrinos from captures on nuclei have mean energies which are significantly less than those produced by capture on protons. Although capture on nuclei under stellar conditions involves excited states in the parent and daughter nuclei, it is mainly the larger $|Q|$-value which significantly shifts the energies of the emitted neutrinos to smaller values. These differences in the neutrino spectra strongly influence neutrino-matter interactions, which scale with the square of the neutrino energy and are essential for collapse simulations[11]. The effects of this more realistic implementation of electron capture on heavy nuclei have been evaluated in independent self-consistent neutrino radiation hydrodynamics simulations by the Oak Ridge and Garching collaborations[12]. Both collapse simulations yield qualitatively the same results and the changes are compared to the previous simulations, which adopted the IPM rate estimate neglecting electron capture on nuclei. In denser regions, the additional electron capture on heavy nuclei results in more electron capture in the new models. In lower density regions, where nuclei with A < 65 dominate, the shell model rates [223] result in less electron capture. The results of these competing effects can be seen in the first

---

[11]A. Mezzacappa et al.: Phys. Rev. Lett. 86, 1935 (2001)

[12]W.R. Hix et al.: Phys. Rev. Lett. 91, 201102 (2003)

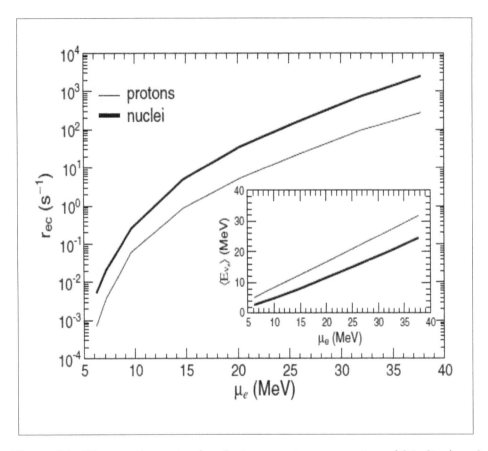

Figure 6.5: The reaction rates for electron capture on protons (thin line) and nuclei (thick line) are compared as a function of electron chemical potential along a stellar collapse trajectory. The insert shows the related average energy of the neutrinos emitted by capture on nuclei and protons. The results for nuclei are averaged over the full nuclear composition (see text). Neutrino blocking of the phase space is not included in the calculation of the rates.

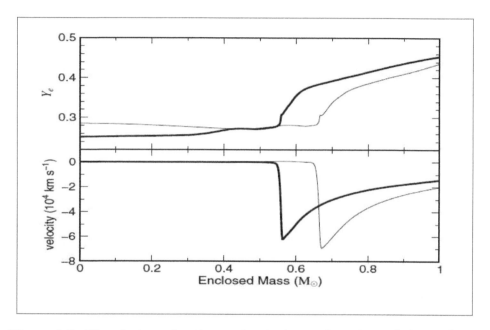

Figure 6.6: The electron fraction and velocity as functions of the enclosed mass at bounce for a $15M_\odot$ model. The thin line is a simulation using the Bruenn parameterization while the thick line is for a simulation using the combined LMP and SMMC+RPA rate sets. Both models employed Newtonian gravity for calculation. Ref. Phys. Rev. Lett. 86, 1678 (2001)

panel of Figure 6.6, which shows the distribution of Y_e throughout the core at bounce (when the maximum central density is reached). The combination of increased electron capture in the interior with reduced electron capture in the outer regions causes the shock to form with 16% less mass interior to it and a 10% smaller velocity difference across the shock. This leads to a smaller mass of the homologous core (by about $0.1 M_\odot$). In spite of this mass reduction, the radius from which the shock is launched is actually displaced slightly outwards to 15.7 km from 14.8 km in the old models. If the only effect of the improvement in the treatment of electron capture on nuclei were to launch a weaker shock with more of the iron core overlying it, this improvement would seem to make a successful explosion more difficult. However, the altered gradients in density and lepton fraction also play an important role in the behavior of the shock. Though also the new models fail to produce explosions in the spherical symmetric limit, the altered gradients allow the shock in the case with improved capture rates to reach 205 km, which is about 10 km further out than in the old models. These calculations clearly show that the many neutron-rich nuclei which dominate the nuclear composition throughout the collapse of a massive star also dominate the rate of electron capture. Astrophysics simulations have demonstrated that these rates have a strong impact on the core collapse trajectory and the properties of the core at bounce. The evaluation of the rates has to rely on theory as a direct experimental determination of the rates for the relevant stellar conditions (i.e. rather high temperatures) is currently impossible. Nevertheless it is important to experimentally explore the configuration mixing between pf and sdg shell in extremely neutron-rich nuclei as such understanding will guide and severely constrain nuclear models. Such guidance is expected from future radioactive ion-beam facilities.

6.5   Neutrino-Induced Processes During a Supernova Collapse

While the neutrinos can leave the star unhindered during the presupernova evolution, neutrino-induced reactions become more and more important during the subsequent collapse stage due to the increasing matter density and neutrino energies; the latter are of order a few MeV in the presupernova models, but increase roughly approximately to the electron chemical potential. Elastic neutrino scattering off nuclei and inelastic scattering on electrons are the two most important neutrino-induced reactions during the collapse. The first reaction randomizes the neutrino paths out of the core and, at densities of a few $10^{11}gcm^{-3}$, the neutrino diffusion time-scale gets larger than the collapse time; the neutrinos are trapped in the core for the rest of the contraction. Inelastic scattering of electrons thermalizes the trapped neutrinos fast within matter and the core collapses as a homologous unit until it reaches densities slightly in excess of nuclear matter, generating a bounce and launching a shock wave which traverses through the in falling material on top of the homologous core. In the currently favored explosion model, the shock wave is not energetic enough to explode the star, it gets stalled before reaching the outer edge of the iron core, but is then eventually revived due to energy transfer by neutrinos from the cooling remnant in the center to the matter behind the stalled shock. The trapped ν_e neutrinos will be released from the core in a brief burst shortly after bounce. These neutrinos can interact with the in falling matter just before arrival of the shock and eventually preheat the matter requiring less energy from the shock for dissociation. The relevant preheating processes are charged- and neutral-current reactions on nuclei in the iron and also silicon mass range. So far, no detailed collapse simulation including preheating has been performed. The relevant cross sections can be calculated on the basis of shell model calculations for the allowed transitions and RPA studies for the forbidden transitions. The main energy transfer to the matter behind the shock, however, is due to neutrino absorption on free nucleons. The efficiency of this transport depends strongly on the neutrino opacities in hot and very dense neutron rich matter. It is likely also supported by convective motion, requiring multidimensional simulations. Finally, elastic neutrino scattering off nucleons, mainly neutrons, also influences the efficiency of the energy transfer. A non-vanishing strange axial vector form factor of the nucleon will likely reduce the elastic neutrino-neutron cross section.

6.6 Type II Supernovae Nucleosynthesis

From the previous discussions it can be said that, a number of aspects of the explosion mechanism are still uncertain and depend on Fe-core sizes from stellar evolution, electron capture rates, the supra nuclear equation of state, the details of neutrino transport and Newtonian vs. general relativistic calculations, as well as multi-D effects[13]. The present situation in supernova modeling is that self-consistent spherically-symmetric calculations do not yield successful explosions based on neutrino energy deposition from the hot collapsed central core (neutron star) into the adjacent layers. Even improve-

---

[13]M. Rampp, H.T. Janka: Astrophys. J. 539, L33 (2000)

ments in neutrino transport, solving the full Boltzmann transport equation for all neutrino flavors, and a fully general relativistic treatment did not change this situation as shown in Figure 6.7. Multi- D calculations, which still have to verify the quality of their neutrino transport schemes did not yet consider the possible combined action of rotation and magnetic fields. The hope that the neutrino driven explosion mechanism could still succeed is based on uncertainties which affect neutrino luminosities (neutrino opacities with nucleons and nuclei, convection in the hot protoneutron star, as well as the efficiency of neutrino energy deposition (convection in the adjacent layers). Figure 6.8 shows the temporal Y_e and entropy evolution of the innermost zone of a $20M_\odot$ SN II simulation with varied neutrino opacities (reducing electron neutrino and antineutron scattering cross sections on nucleons or increasing the neutrino absorption cross sections for $\nu_e + n \rightarrow p + e^-$ and $\overline{\nu_e} + p \rightarrow n + e^+$). Such parameter studies permit successful delayed explosions and show options which may lead to success. This tests variations in the uncertainty of neutrino cross sections but probably beyond the permitted range as factors of about 2 are involved. Such tests can, however, also mimic multi-D effects in a spherically symmetric calculation. Decreased neutrino scattering cross sections cause a larger neutrino luminosity, similar to enhanced neutrino transport via protoneutron star convection. Larger neutrino absorption cross sections act like an increase in energy absorption efficiency, similar to convection beyond the neutrino sphere. The interesting feature of the results is that in successful cases a zone with $Y_e > 0.5$ exists in the innermost ejecta (caused by neutrino interactions, shown in Figure 6.8), leading to a proton-rich and alpha-rich freeze-out of explosive Si-burning with high entropies. This leads similar to the big bang, but on much more minute scales, to unburned He and H in the innermost ejecta shown in Figure 6.9. As observations show typical kinetic energies of $10^{51} erg$ in supernova remnants, in the past light curve as well as explosive nucleosynthesis calculations were performed, even without a correct understanding of the explosion mechanism. They artificially introduced a shock of appropriate energy in the pre-collapse stellar model [14] and followed the explosive nucleosynthesis caused by the shock front passing through the layers up to the stellar surface. Such calculations lack self-consistency and cannot predict the ejected ^{56}Ni-masses from the innermost explosive Si-burning layers (powering supernova light curves by the decay chain ^{56}Ni-^{56}Co-^{56}Fe) due to missing knowledge about the detailed explosion mechanism and therefore the mass cut between the neutron star and supernova ejecta. However, the intermediate mass

---

[14]T. Rauscher, A. Heger, R.D. Hoffman et al.: Astrophys. J. 576, 323 (2002)

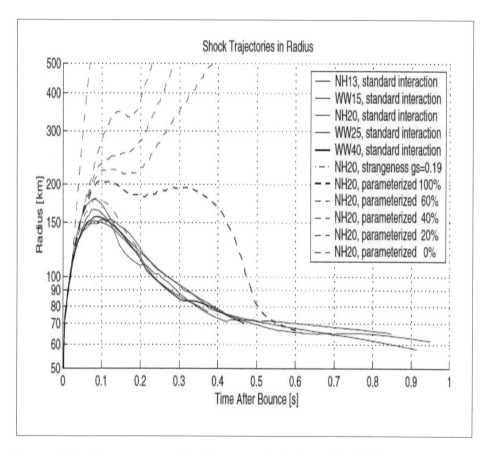

Figure 6.7: A sequence of collapse calculations for different progenitor masses, showing in each case the radial position of the shock front after bounce. We see that the shocks are strongest for the least massive stars. But in these 1D calculations all of them stall, recede and turn into indexaccretionaccretion shocks, i.e. not causing successful supernova explosions. A reduction in neutrino-nucleon elastic scattering, leading to higher luminosities, can cause explosions (performed for a $20M_\odot$ star). Ref: Phys. Rev. D 63, 4003(2001)

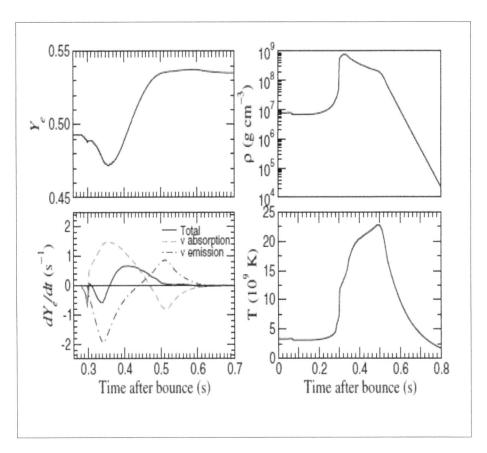

Figure 6.8: Hydrodynamic simulations with varied (reduced) neutrino opacities (scattering cross sections on nucleons reduced by 60%) lead to larger neutrino luminosities and make successful supernova explosions possible. Here we see the time evolution after core bounce of the innermost ejected layer from a $20M_\odot$ supernova progenitor. $Y_e = <Z/A>)$ indicates the neutron-richness of ejected matter, $\frac{dY_e}{dt}$ the time derivative due to the different reactions involving free protons and neutrons $\nu_e + n \rightleftharpoons p + e^-$ and $\overline{\nu_e} + p \rightleftharpoons n + e^+$ in the directions of neutrino (and antineutrino) absorption or emission. $\rho(t)$ and $T(t)$ the indicate density and temperature. What can be noticed is that Y_e is strongly dependent on both neutrino absorption and emission reactions and that apparently in these exploding models Y_e in the innermost zones is larger than 0.5, i.e. proton-rich.

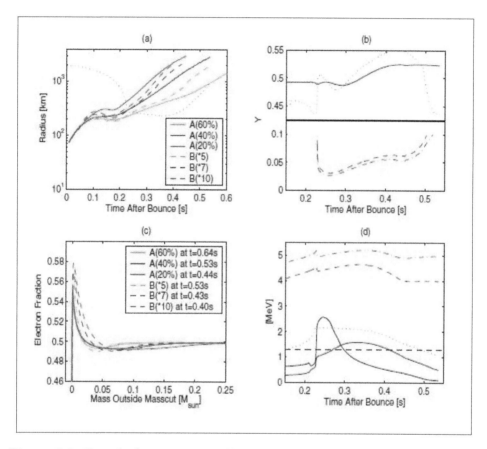

Figure 6.9: Details from a core collapse and explosion simulation with six variations in neutrino transport, (A) reduction of neutrino-nucleon scattering cross sections to a given percentage of the literature value, (B) multiplication of the neutrino and antineutrino absorption cross sections on nucleons by a given factor. Shown are (a) the radius of the shock front for the different models, (b) the electron fraction Y_e, i.e. the net electron (or proton) to nucleon ratio, for the innermost ejected mass zone of one simulation as a function of time in comparison to a value where neutrino emission and absorption rates are in a chemical equilibrium, (c) Y_e, measuring the neutron or proton-richness of matter, for a number of ejected mass zones outside the mass cut, (d) the neutrino chemical potential and the matter temperature.

elements Si-Ca are only dependent on the explosion energy and the stellar structure of the progenitor star, while abundances for elements like O and Mg are essentially determined by the stellar progenitor evolution. Thus, when moving in from the outermost to the innermost ejecta of a SN II explosion, we see an increase in the complexity of our understanding, depending (a) only on stellar evolution, (b) on stellar evolution and explosion energy, and (c) on stellar evolution and the complete explosion mechanism (see Figure 6.10). The possible complexity of the explosion mechanism, including multi-D effects, should not affect this (spherically symmetric) discussion of explosive nucleosynthesis severely. 2D-calculations led to spherically symmetric shock fronts after the explosion is initiated, thus causing spherical symmetry in explosive nuclear burning when passing through the stellar layers. Only after the passage of the shock front, the related temperatures decline and freeze-out of nuclear reactions, the final nucleosynthesis products can be distributed in non-spherical geometries due to mixing by hydrodynamic instabilities. Thus, the total mass of nucleosynthesis yields shown in Figure 6.10 should not be changed, only its geometric distribution. An exception is probably related to the very innermost ejected zones where the explosion mechanism is influencing the outcome. The correct prediction of the amount of Fe-group nuclei ejected (which includes also one of the so-called alpha elements, i.e. Ti) and their relative composition depends directly on the explosion mechanism and the size of the collapsing Fe-core. Three types of uncertainties are inherent in the Fe-group ejecta, related to (i) the total amount of Fe (group) nuclei ejected and the mass cut between neutron star and ejecta, mostly measured by ^{56}Ni decaying to ^{56}Fe, (ii) the total explosion energy which influences the entropy of the ejecta and with it the amount of radioactive ^{44}Ti as well as ^{48}Cr, the latter decaying later to ^{48}Ti and being responsible for elemental Ti, and (iii) finally the neutron richness or $Y_e = <Z/A>$ of the ejecta, dependent on stellar structure, electron captures and neutrino interactions. Y_e influences strongly the ratios of isotopes 57/56 in Ni(Co,Fe) and the overall elemental Ni/Fe ratio, the latter being dominated by ^{58}Ni and ^{56}Fe. In the inner ejecta, corresponding to case (c) in the discussion above, which experience explosive Si-burning, such (not self-consistent) calculations made use of a Y_e of the order 0.4989 to 0.494. This omitted possible alterations of Y_e via neutrino reactions during the explosion. This can cause huge changes in the Fe-group composition [74,237] for 56Ni and the neutron-rich isotopes $^{57}Ni,^{58}Ni,^{59}Cu,^{61}Zn$, and ^{62}Zn. The abundances of $^{40}Ca,^{44}Ti,^{48}Cr$, and ^{52}Fe are affected by the entropy obtained in

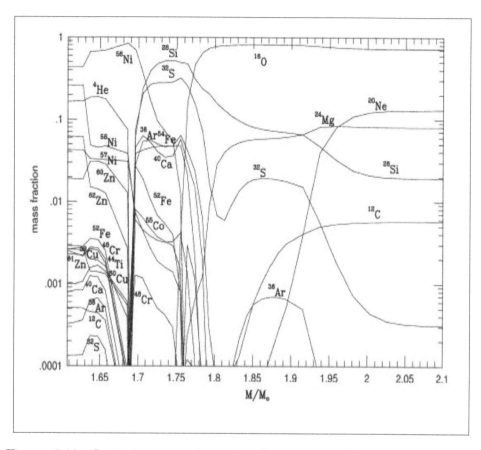

Figure 6.10: Isotopic composition for the explosive C-, Ne-, O- and Si-burning layers of a core collapse supernova from a $20M_\odot$ progenitor star with a $6M_\odot$ He-core and an induced net explosion energy of $10^{51}erg$, remaining in kinetic energy of the ejecta. M(r) indicates the radially enclosed mass, integrated from the stellar center. The exact mass cut in M(r) between neutron star and ejecta and the entropy and Y_e in the innermost ejected layers depend on the details of the (still open) explosion mechanism. The abundances of O, Ne, Mg, Si, S, Ar, and Ca dominate strongly over Fe (decay product of ^{56}Ni), if the mass cut is adjusted to $0.07M_\odot$ of ^{56}Ni ejecta as observed in SN 1987A. Ref: Astrophys. J. 460, 408 (1996)

those mass zones. Calculations by Liebendörfer et al[15] show that apparently the innermost zones in successfully exploding models avoid too neutron-rich ejecta when including neutrino-induced reactions. This permits an Fe-group composition consistent with supernova observations. The slightly proton-rich environment leads to an alpha-rich and proton-rich freeze-out. The density is apparently not sufficient for an rp-process and unburned hydrogen survives deep in the ejected envelope. The pending understanding of the explosion mechanism affects the amount of Fe-group nuclei ejected (which includes also one of the so-called alpha elements, i.e. Ti) and possible r-process yields for SNe II (see Chapter 7). The pending understanding of the explosion mechanism also affects possible r-process yields for SNe II as described in detail in Chapter 7.

Exercise

1. A neutron star is a compact, dense object made of degenerate neutrons having a density similar to that in the central part of a heavy nucleus. (a) If the density of nuclear matter is $0.17 nucleons/fm^3$ or $2.8 \times 10^{17} kg/m^3$, what is the radius of a neutron star having a mass one and a half times that of the sun? (One solar mass, $M_\odot = 2.0 \times 10^{30} kg$.)

   (b) A neutron star is one of the possible remnants of a supernova explosion such as SN 1987a, the one which took place in the Large Magellanic Cloud 160,000 light years away and was first observed on earth on February 24, 1987. When the core of a large star exhausts its nuclear fuel, there is no longer the thermal pressure to counterbalance the gravitational force, and the core of the star collapses. For simplicity, we can consider that all the material in the core of the collapsing star is in the form of ^{56}Ni made of 28 neutrons and 28 protons. Because of the tremendous gravitational force, the protons in ^{56}Ni change into neutrons by capturing atomic electrons through the reaction $p + e^- \rightarrow n + \nu_e$, Calculate the number of neutrinos released in converting 1.5 solar mass of ^{56}Ni atoms into neutrons during the gravitational collapse.

   (c) If the total cross section for a neutrino to interact with each nucleon is $10^{-48} m^2$, how many reactions due to the neutrinos from such a gravitational collapse can one expect in a detector on earth made of 3000 tons of water? Compare this with the number of events observed with such a detector at Kamioka due to supernova SN 1987a.

   (d) Assuming that the average energy of each neutrino is 10MeV in

---

[15]M. Liebendörfer, A. Mezzacappa, O.E.B. Messer et al.: Nucl. Phys. A 719, 144 (2003)

such an event, calculate the total amount of energy carried away by the neutrinos from the gravitational collapse. Compare this value with the rest-mass energy of the sun.

2. What is the gravitational energy released when non-interacting particles with a total mass $1.99 \times 10^{30} kg$ collapse from an infinite distance of separation to a spherical ball of radius $6.96 \times 10^{8} m$? Assuming an ideal gas equation of state, what is the temperature of this sphere? Compare the result with the value of $15 \times 10^{6} K$ for the interior of our sun.

References

Quoted in the text:

- F.C. Barker, T. Kajino: Aust. J. Phys. 44, 369 (1991)

- R. Azuma et al.: Phys. Rev. C 50, 1194 (1994)

- L. Buchmann: Astrophys. J. 468, L127 (1996); Ap. J. 479, L153 (1997)

- L. Buchmann, R.E. Azuma, C.A. Barnes, J. Humblet, K. Langanke: Phys. Rev. C 54, 393 (1996); Nucl. Phys. A 621, 153 (1997)

- L. Buchmann et al.: Phys. Rev. Lett. 70, 726 (1993)

- G.R. Caughlan, W.A. Fowler: At. Data Nucl. Data Tables 40, 283 (1988)

- G.R. Caughlan, W.A. Fowler, M.J. Harris, G.E. Zimmerman: At. Data Nucl. Data Tables 32, 197 (1985)

- B.W. Filippone, J. Humblet, K. Langanke: Phys. Rev. C 40, 515 (1989)

- M. Fey et al.: Nucl. Phys. A 718, 131

- A. Heger, K. Langanke, G. Martinez-Pinedo, S.E. Woosley: Phys. Rev. Lett. 86, 1678 (2001)

- A. Heger, S.E. Woosley, G. Martinez-Pinedo, K. Langanke: Astrophys. J. 560, 307 (2001)

Chapter 7

Basic Components of Astrophysical r-Process

The heavy elements found in nature are manufactured by neutron capture processes occurred in two types of different environments viz. (i) A process with small neutron densities, experiencing long neutron capture timescales in comparison to β-decays ($\tau_\beta < \tau_{n,\gamma}$, slow neutron capture or the s-process, causing abundance peaks in the flow path at nuclei with small neutron capture cross sections, i.e. stable nuclei with closed shells at magic neutron numbers as a consequence of (α, n)-reactions in hydrostatic He-burning. (ii) A process with high neutron densities and temperatures, experiencing rapid neutron captures and the reverse photo disintegrations with $\tau_{n,\gamma}, \tau_{\gamma,n} < \tau_\beta$, causing abundance peaks due to long β-decay half-lives where the flow path comes closest to stability (again at magic neutron numbers, but for unstable nuclei). During the latter rapid neutron-capture process (r-process) it is possible that highly unstable nuclei with short half-lives are produced, leading also to the formation of the heaviest elements in nature like Th, U, and Pu. The r-process path involves nuclei near the neutron drip-line[1]. Far from stability, neutron shell closures are encountered for smaller mass numbers A than in the valley of stability. Therefore, if r-process peaks are due to long β-decay half-lives of neutron magic nuclei, the r-process abundance peaks are shifted in comparison to the s-process peaks (which occur for neutron shell closures at the stability line). Besides this basic understanding, the history of r-process research has been quite diverse in suggested scenarios.

---

[1]In a plot of Z versus N, the limiting line along which single neutron separation energy is zero is called the neutron drip-line. And the limiting line along which single proton separation energy is zero is called the proton drip-line

If starting with a seed distribution somewhere around A=50-80 before rapid neutron-capture sets in, the operation of an r-process requires 10 to 150 neutrons per seed nucleus to form all heavier r-nuclei. The question is which kind of environment can provide such a supply of neutrons to act before decaying with a 10 min half-life. The logical conclusion is that only explosive environments, producing or releasing these neutrons suddenly, can account for such conditions.

7.1 Role of Nuclear Physics in the r-process

The main aspects of an r-process are neutron captures, photodisintegrations, and β-decays. An $(n, \gamma) \rightleftharpoons (\gamma, n)$ equilibrium exists if neutron captures and photodisintegrations are fast in comparison to beta-decays between isotopic chains, leading to a distribution of abundances in each isotopic chain governed by a chemical equilibrium $\mu_n + \mu_{Z,A} = \mu_{Z,A+1}$ in a Boltzmann gas. This causes abundance ratios of neighboring isotopes $Y(Z, A + 1)/Y(Z, A) = f(n_n, T, S_n)$ to depend only on neutron density n_n, temperature T, and the neutron separation energy S_n (or reaction Q-value for the appropriate neutron capture). The maximum in each isotopic chain occurs when $Y(Z, A + 1)/Y(Z, A)$ changes from a rising to a declining ratio (i.e. from > 1 to < 1). A ratio of 1 defines a universal $S_{n,max}$ for all maximums, independent of the specific isotopic chain. Thus, the combination of a neutron density n_n and temperature T determines the r-process path (connecting the isotopes with the maximum abundance in each isotopic chain) located at the same neutron separation energy $S_{n,max}$. During an r-process event exotic nuclei with neutron separation energies of 4 MeV and less are important, up to $S_n = 0$, i.e. the neutron drip-line. This underlines that the understanding of nuclear physics far from stability is a key ingredient and the knowledge of S_n (or equivalently nuclear masses) determines the r-process path. While in recent years new Hartree-Fock(-Bogoliubov) or relativistic mean field approaches have been addressing this question, the FRDM model still seems to provide the best reproduction of known masses. The β-decay rates λ_Z, A_β are related to the half-lives of very neutron-rich nuclei via $\lambda_\beta = ln(2)/\tau_{1/2}$. The abundance flow from each isotopic chain to the next is governed by β-decays. We can introduce a total abundance in each isotopic chain $Y(Z) = \sum_A Y(Z, A)$. Process timescales in excess of β- decay half-lives lead to a steady-flow equilibrium $Y(Z)\lambda_\beta(Z) = $ const. Thus, in that case the knowledge of nuclear masses (S_n), determining the r-process path, and half-lives ($\tau_{1/2}$), determining the relative abundances of each isotopic chain, would be sufficient to predict the whole set of abundances. This seems to be the case in the regions between the r-process peaks (neutron magic numbers) and in the low-mass tails of the $A \approx 130$ and $A \approx 195$ peaks of the solar-system r-process abundances. Nuclei in the r-process path with the longest half-lives of the order 0.2-0.3s, related to the abundance peaks themselves, do not fulfill the steady flow requirement. In this coupled set of differential equations

$$\dot{Y}(Z) = \lambda_\beta(Z-1)Y(Z-1) - \lambda_\beta(Z)Y(Z) \qquad (7.1)$$

has to be solved for all isotopic chains Z, if an $(n,\gamma) \rightleftharpoons (\gamma,n)$ equilibrium applies. In the most general case of (astrophysical) environment conditions, one has to solve a full set of differential equations for all nuclei from stability to the neutron drip-line, including individual neutron captures, photodisintegrations, and beta-decays. However, from existing results one finds that an $(n,\gamma) \rightleftharpoons (\gamma,n)$ equilibrium is attained before the freeze-out of neutron abundances and photodisintegrations (for decreasing temperatures). For small β-decay half-lives, encountered in between magic numbers and for small Z_s at magic numbers, also the steady-flow approximation seems applicable. The freeze-out from equilibrium can follow two extreme options: (i) an instantaneous freeze-out, just followed by the final decay back to stability, where also β-delayed properties (neutron emission and fission) are needed and can depend strongly on the beta strength-function[2]. (ii) In the more general case of a slow freeze-out also neutron captures can still affect the final abundance pattern[3]. Fission will set in during an r-process, when neutron-rich nuclei are produced at excitation energies beyond their fission barriers. The role of β-delayed and neutron-induced fission has two aspects. For nuclei with neutron separation energies of the order 2 MeV, neutron capture will produce compound nuclei with much smaller excitation energies than those obtained in β-decay. However, the rates of neutron-induced processes (responsible also for the $((n,\gamma) \rightleftharpoons (\gamma,n)$ equilibrium) are orders of magnitude larger than beta decay rates. Thus, it is possible that neutron-induced fission can compete with beta-delayed fission (see Figure 7.1). Fission determines on the one hand the heaviest nuclei produced in an r-process and on the other hand also the fission yields fed back to lighter nuclei. In supernovae, a high neutrino flux of different flavors is available. This gives rise to neutral and charged current interactions with nucleons and

---

[2]F.-K. Thielemann et al.: In: Fifty Years with Nuclear Fission, ed. J.W.Behrens, A.D. Carlson (American Nuclear Society,1989) p. 592

[3]R. Surman et al.: Phys. Rev. Lett. 79, 1809 (1997)

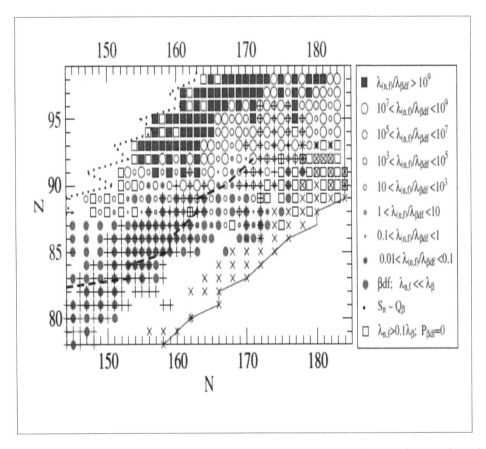

Figure 7.1: Ratios of neutron-induced to beta-delayed fission for nuclei of interest in r-process calculations. These exploratory results were obtained with the fission barriers by Howard and Möoller(At. Data Nucl. Data Tables 25, 219 (1980)). Different barrier heights will leads to different total fission rates. However, it was shown here that neutron-induced fission is a viable ingredient for r-process studies and should not be neglected in comparison to beta-delayed fission. Ref:Astronomy Letters 29, 510 (2003)

nuclei, i.e. elastic/inelastic scattering or electron neutrino or antineutrino capture on nuclei, e.g. $\nu_e + (Z, A) \rightarrow (Z + 1, A) + e^-$, (giving results similar to β transformations). During freeze-out the first mechanism redistributes abundances to nearby mass numbers similar to spallation. Neutrino capture for high neutrino fluxes could mimic fast β-decays, possibly accelerating an r-process to heavy elements. Site-independent classical analyses, based on neutron number density n_n, temperature T, and duration time τ, as well as entropy based calculations with the parameters entropy S, Y_e, and expansion timescale τ have shown that the solar r-process can be fitted by a continuous superposition of components with neutron separation energies at freeze-out in the range 4-1 MeV[4]. These are the regions of the nuclear chart (extending to the neutron drip-line) where nuclear structure, related to masses far from stability and beta-decay half-lives has to be investigated. They include predominantly nuclei not accessible in laboratory experiments to date but hopefully in the foreseeable future (RIKEN, GSI, RIA). In addition, expanding theoretical efforts with increasingly sophisticated means are required, addressing also neutrino-induced reactions and fission properties. The actual astrophysical realization of the relevant conditions will be discussed in the following sections.

7.2 Parameters controlling the r-process

The parameter which determines whether an r-process occurs is the neutrons per seed ratio which is 10 to 150 neutrons per r-process seed in the Fe-peak or somewhat beyond, permitting to produce nuclei with A > 200. This can be translated into the parameters entropy S, Y_e and expansion timescale τ of a heated blob of material in astrophysical events, consisting of a net ratio of protons to neutrons (or total nucleons) with a given expansion history. At low entropies (without an alpha-rich freeze-out of charged-particle reactions) this is equivalent to $Y_e = <Z/A> = 0.12 - 0.3$. Such a high neutron excess is only possible for high densities in neutron stars under beta equilibrium ($e^- + p \leftrightarrow n + \nu, \mu_e + \mu_p = \mu_n$), based on the high electron Fermi energies which are comparable to the neutron-proton mass difference. Deviations from this straightforward balance are only possible if one stores large amounts of mass in N=Z nuclei with small neutron capture cross sections (e.g. ^{4}He), leaving then all remaining neutrons for a few heavy seed nuclei. This phenomenon is known as an extremely α-rich freeze-out in complete Si-burning and corresponds to a weak link of reactions between the light nuclei

---

[4]J.J. Cowan et al.: Astrophys. J. 521, 194 (1999)

(n, p, α) and heavier nuclei at low densities. The links across the particle-unstable A=5 and 8 gaps are only possible via the three-body reactions $\alpha\alpha\alpha$ and $\alpha\alpha n$ to ^{12}C and ^{9}Be, whose reaction rates show a quadratic density dependence. The entropy ($\propto T_3/\rho$ in radiation dominated matter) can be used as a measure of the ratio between the remaining ^{4}He mass fraction and heavy nuclei. A well known case is the big bang where under extreme entropies essentially only ^{4}He is left as the heaviest nucleus available. Somewhat lower entropies permit the production of (still small) amounts of heavy seed nuclei. Then, even moderate values of $Y_e = 0.4 - 0.5$ can lead to high ratios of neutrons to heavy nuclei for entropies in excess of 200 kobo per baryon. And neutron captures can proceed to form the heaviest r-process nuclei. These two environments represent a normal (low-entropy) and an α-rich (high-entropy) freeze-out from charged-particle reactions before the dominance of neutron-induced reactions. Towards low entropies the transition to a normal freeze-out occurs, leading to a negligible entropy dependence of the neutron to seed ratio.

7.3 r-Process Sites

7.3.1 Type II Supernovae.

The apparently most promising mechanism for supernova explosions after Fe-core collapse of a massive star is based on neutrino heating beyond the hot proto-neutron star via the dominant processes $\nu_e + n \rightarrow p + e^-$ and $\overline{\nu_e} + p \rightarrow n + e^+$ with about 1% efficiency in energy deposition. The neutrino heating efficiency depends on the neutrino luminosity, which in turn is affected by neutrino opacities. Aspects of the explosion mechanism are still uncertain. If SNe II are also responsible for the solar r-process abundances, given the galactic occurrence frequency, they would need to eject about $10^{-5}M_\odot$ of r-process elements per event (if all SNe II contribute equally). The scenario is based on the so-called "neutrino wind", i.e. a wind of matter from the neutron star surface (within seconds after a successful supernova explosion) caused by neutrinos streaming out from the hot neutron star. This high entropy neutrino wind is expected to lead to a superposition of ejecta with varying entropies. If a sufficiently high entropy range is available, an abundance pattern as shown in Figure 7.2 can be obtained. However, the r-process by neutrino wind ejecta of SNe II faces two difficulties. (i) Whether the required high entropies for reproducing heavy r-process nuclei can really be attained in supernova explosions has still to be verified. Because it appears that only large or compact neutron stars with masses in excess of $2M_\odot$

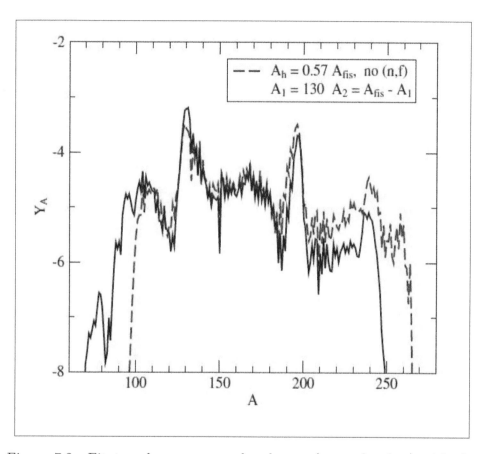

Figure 7.2: Fit to solar r-process abundances (open rhombus) with the ETFSI-1 mass formula (Ref: Y. Aboussir et al.: At. Data Nucl. Data Tables 61, 127 (1995)), making use of a superposition of entropies. These calculations were performed with $Y_e = 0.45$, but similar results are obtained in the range 0.30-0.49, only requiring a scaling of entropy. The trough below $A \simeq 130$ can be avoided by the changing of shell effects far from stability. The strong deficiencies in the abundance pattern below $A \simeq 110$ are due to an α-rich freeze-out where essentially no neutrons are left. This is related to the Y_e of the astrophysical scenario rather than to nuclear structure (Ref. C. Freiburghaus et al.: Astrophys. J. 516, 381 (1999).

can provide the high entropies required. (ii) The mass region 80 -110 experiences difficulties to be reproduced adequately, reflecting rather abundances determined by alpha separation energies after an alpha-rich freeze-out than neutron separation energies. It has to be seen whether the inclusion of non-standard neutrino properties can cure both difficulties or lower Y_e zones can be ejected from SNe II, as claimed by Sumiyoshi et al.[5] from assumed prompt explosion calculations, lacking a proper neutrino transport. Present supernova models face the problem that the entropies required seem not yet attainable, unless ad hoc assumptions can be verified.

7.3.2 Neutron Star Mergers

An alternative site for the heavy r-process nuclei are neutron-star ejecta, like e.g. in neutron star mergers[6]. The binary system, consisting of two neutron stars, looses energy and angular momentum through the emission of gravitational waves and merges finally. The measured orbital decay gave the first evidence for the existence of gravitational radiation[7] and indicates timescales of the order of 10^8 y or less depending on the eccentricity of the system. The rate of NS mergers has been estimated to be of the order $10^{-6} - 10^{-4} y^{-1}$ per galaxy. A merger of two NS can also lead to the ejection of neutron-rich material of the order of $10^{-(2-3)} M_{\odot}$ in Newtonian and relativistic calculations[8]. Although, the decompression of cold neutron-star matter has been studied, a hydro dynamical calculation coupled with a complete r-process calculation has yet not been undertaken. Figure 7.3 shows the composition of ejecta from a NS merger. It is seen that the large amount of free neutrons (up to $n_n \simeq 10^{32} cm^{-3}$) available in such a scenario leads to the build-up of the heaviest elements and also to fission cycling within very short timescales, while the flow from the Fe-group to heavier elements "dries up". This leads to a composition void of abundances below the $A \simeq 130$ peak, which is, however, dependent on detailed fission yield predictions.

7.3.3 r-Process Overview

Presently, the suggested r-process sources, supernovae and neutron star mergers, did not yet prove to be "the" main-agent of r-process source without arguable doubt. A discussion of the advantages and disadvantages of

[5]K. Sumiyoshi et al.: Astrophys. J. 562, 880 (2001)

[6]J.M. Lattimer et al.: Astrophys. J. 213, 225 (1977)

[7]J. Taylor: Rev. Mod. Phys. 66, 711 (1994)

[8]R. Oechslin et al.: Phys. Rev. D 65, 3005 (2002)

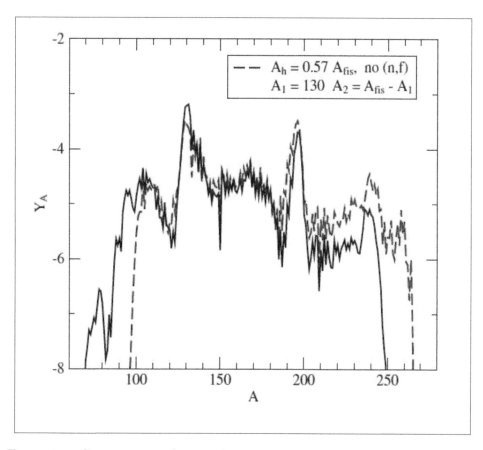

Figure 7.3: Composition of ejecta from a neutron star merger event (Σ mass fractions $=1$) with an average $Y_e = 0.10$. Either only beta-delayed fission or both beta-delayed and neutron-induced fission were employed, varying also the fission yield distribution. The effect of these different treatments is obviously seen for A$>$240, but more importantly for A$<$130 and underlines the impact of fission barriers and the fission yield distribution. These effects are directly related to all events with very high neutron densities, i.e. neutron star ejecta in mergers or jets where strong fission cycling takes place.[ref:Astronomy Letters 29, 510 (2003)]

both possible r-process sources (SNe II vs. neutron star mergers) has been presented by Qian[9]. Although self-consistent core collapse supernovae do not give explosions, yet, parameter studies with neutrino opacities permit to "fit" the correct explosion behavior. Thus, there is no way to predict whether the required entropies for an r-process can be obtained. Neutron star merger calculations give the correct mass ejection, but relativistic calculations need to be followed up and Y_e needs to be treated with weak interactions and neutrino transport included self-consistently. The two possible sites discussed above (SNe II and neutron star mergers) have different occurrence frequencies and different amounts of r-process ejecta, if a successful r-process actually occurs. These properties will enter into the enrichment pattern of r-process elements in galactic evolution. Inhomogeneous galactic evolution models in the very early phases of the Galaxy will finally indicate that either one of the above mentioned sites or both in combination can meet the observational constraints from the scatter of r-process to Fe ratios in "low metallicity stars" as a function of metallicity, i.e. the $(Fe/H)/(Fe/H)_\odot$ ratio. Nuclear properties far from stable nuclei are of prime importance in the astrophysical r-process. Due to the (partial) equilibrium nature at high temperatures and neutron densities, the dominant influence is given by nuclear masses and β-decay properties (half-lives, delayed neutron emission and fission). Neutron-induced fission can play a competing role to beta-delayed fission. All these aspects need a better future understanding, including the role of neutrino-induced reactions, as they enter directly in the resulting r-process yields and abundance distributions.

Exercise

1. What are meant by the astrophysical r-process? What is its role in nuclear astrophysics?

2. Discuss the role of nuclear physics in the astrophysical r-process.

3. Discuss the parameters that conrtol the r-processes.

4. Give short note of type-II supernovae.

5. What is nuetron star? What are netron star mergers? Make an estimate of the approximate radius of a typical neutron star.

---

[9]Y.-Z. Qian: Astrophys. J. 534, L67 (2000)

References

Quoted in text:

- H.E. Suess, H.C. Urey: Rev. Mod. Phys. 28 53 (1956)

- P.A. Seeger et al.: Astrophys. J. Suppl. 97, 121 (1965)

- G. Wallerstein et al.: Rev. Mod. Phys. 69, 995 (1997)

- S. Wanajo et al.: Astrophys. J. 554, 578 (2001)

- J.J. Cowan et al.: Phys. Rep. 208, 267 (1991)

- E.M. Burbidge et al.: Rev. Mod. Phys. 29, 547 (1957)

- A.G.W. Cameron: Atomic Energy of Canada, Ltd., CRL-41 (1957)

- C. Freiburghaus et al.: Astrophys. J. 516, 381 (1999)

- C. Freiburghaus et al.: Astrophys. J. 525, L121 (1999)

- J.M. Lattimer et al.: Astrophys. J. 213, 225 (1977)

Chapter 8

Nuclear Processes in Explosive Binary Stars

8.1 Introduction

Over the last twenty years thermonuclear explosions in accreting binary star systems have been an object of considerable attention. The basic concept of the thermonuclear runaway as the driving explosion mechanism seems reasonably well understood but there are still considerable discrepancies between the predicted observables and the actual observations. The proposed mechanism involves binary systems with one (or two) degenerate objects, like white dwarfs or neutron stars and is characterized by the revival of the dormant objects via mass overflow and indexaccretionaccretion from the binary companion. This leads to explosive events like novae, type Ia supernovae, X-ray bursts, and X-ray pulsars. The characteristic differences in the luminosity, time scale, and periodicity depend on the indexaccretionaccretion rate and on the nature of the accreting object. Low indexaccretionaccretion rates lead to a pile-up of unburned hydrogen, causing the ignition of hydrogen burning via pp-chains and CNO-cycles with pycnonuclear enhancements of the reactions after a critical mass layer is attained. On white dwarfs this triggers nova events, on neutron stars it results in X-ray bursts. High indexaccretionaccretion rates above a critical limit on the other hand cause high temperatures in the accreted envelope and less degenerate conditions, which result in stable H-burning or only weak flashes. High indexaccretionaccretion rates on white dwarfs cause supernovae type Ia events, which are not discussed here in detail as they do not lead to nuclei far from stability[1]. Ac-

---

[1]F.-K. Thielemann et al., New Astronomy 48, 65 (2004)

cretion rates high enough for stable burning on a highly magnetic neutron star lead to X-ray pulsars. The present modeling of the accretion and thermonuclear runaway process is still in infancy mainly due to the complex aspects of the explosion mechanism which requires three-dimensional modeling techniques for realistic treatment. Besides the complexities of asymmetric ignition and burning front development and the issues of rapid convection and mixing during the explosion large uncertainties are also associated with the microscopic nuclear physics component of the process. The nuclear energy generation provides the observed luminosity of the event, the combination of rapid mixing, convection and far of stability nucleosynthesis is responsible for the observed abundance distribution in the ejected material. These events are the largest thermonuclear explosions in the universe, they may synthesize a number of important isotopes in our universe, and they serve as laboratories for nuclear physics at extreme temperatures and densities.

8.2 Nova Explosions

Novae are astronomical events where a star suddenly brightens and gives off a large amount of energy for a period of time. Novae are associated with binary stellar systems and are often recurrent. A typical nova outburst occurs when a larger companion star accretes hydrogen-rich matter onto a smaller white dwarf companion. The accreted matter forms a degenerate gas shell on the white dwarf's surface and eventually ignites with hydrogen burning via the CNO cycle. CNO burning burns hydrogen into helium, utilizing a set of reactions where carbon, nitrogen, and oxygen serve as catalysts in a cycle reaching relatively steady state while hydrogen decreases and helium increases. CNO burning in the nova eventually heats the accreted matter to a high enough temperature that the "hot" CNO cycle becomes the predominant source of energy production. Thus novae hcan be interpreted as thermonuclear runaways on the surface in close binary star systems. Accretion processes or mass exchange between the two binary stars can take place when at least one of the stars fills its Roche Lobe[2] which is the gravitational equipotential surface enclosing both stars. The matter of the extended star flows through the Roche Lobe onto the surface of the second companion which represents a deeper gravitational potential. Novae are identified as white dwarfs in binary systems, accreting matter from the binary companion star in a burning stage close to main sequence H-burning which is filling its

---

[2]The Roche lobe is the region around a star in a binary system within which orbiting material is gravitationally bound to that star.

Roche Lobe. The accreted material forms a thin, but high density electron degenerate envelope at the surface of the white dwarf. Dredge up of white dwarf material (^4He, ^{12}C, ^{16}O in the case of an CO-white dwarf, ^{16}O, ^{20}Ne, and ^{24}Mg in the case of an ONeMg-white dwarf) into the envelope leads to an enrichment of the accreted material in heavier isotopes. The exact nature of the mixing mechanism is still under debate but it has recently been suggested that mixing of white dwarf material into the accreted envelope can be sufficiently enhanced by inter facial breaking gravity wave interaction. After a "critical" mass has been accreted, thermonuclear ignition takes place at the bottom of the accreted envelope. This depends critically on the mass of the white dwarf and the accretion rate which determines the pressure conditions at the bottom of the envelope. The ignition presumably occurs via the pp-chains, at degenerate electron gas conditions this causes a rapid increase in temperature at constant pressure and density. This thermonuclear runaway is further enhanced by the subsequent ignition of the hot CNO cycles on the high abundances of ^{12}C and ^{16}O until the degeneracy is lifted after the Fermi temperature T_F has been reached. The Fermi temperature depends on the electron density $n_e \propto \rho Y_e$ and

$$T_F = 3.03 \times 10^5 (\rho Y_e)^{2/3}, \tag{8.1}$$

with ρ in g/cm^3 and T_F in K. Thus, higher densities of the material lead to higher peak temperatures which can be reached in the thermal runaway before degeneracy is lifted. However, if the shell temperature is rising rapidly the peak temperature can exceed the Fermi temperature before the electron gas is sufficiently degenerate to initiate expansion. Due to the rapid temperature increase at the bottom of the envelope a convective zone develops which gradually grows to the surface as the temperature continues to increase. This allows rapid energy transport to the surface within the convective time scale of $t_{conv} \approx 10^2 s$. Within that short timescale also an appreciable fraction of the long-lived β^+ emitters which are produced by the hot CNO cycles is carried to the surface. The release of decay energy further increases the luminosity to values above $10^5 L_\odot$. The large amount of energy deposited in the outer layers on the short convective time scale, coupled with high luminosity often exceeding the Eddington limit[3] L_{edd}[4] causes rapid expansion of the outer layers and the ejection of the outer shells. Typical novae are characterized by thermal runaways with densities of approximately $\rho \approx 10^3 g/cm^3$ and typical peak temperatures between $1 \times 10^8 K$ and $4 \times 10^8 K$. The main observables for a reliable interpretation of the explosion mechanisms are the nova light curve and the abundance distribution in the ejected material. The luminosity yields information on the total energy release from nuclear reaction and decay processes but it also gives information about the time scale of the thermonuclear runaway. The abundance observation gives evidence for the on-site nucleosynthesis but may also provide a tool for monitoring the rather complex mixing and convective mechanisms prior and during the explosion. For a reliable interpretation of such observations improved nuclear physics input in the present nova models is crucial. The actual ignition temperature for novae is well below these peak temperatures. Rates of many nuclear reactions are therefore needed at energies below a few hundred keV to get a complete description of the ignition process in the various accreted layers of material. The main energy generation in nova comes from the hot (or β-limited) CNO cycles $^{12}C(p,\gamma)^{13}N(p,\gamma)^{14}O(\beta,\nu)^{14}N(p,\alpha)^{15}O(\beta,\nu)^{15}N(p,\alpha)^{12}C$ and is determined by the rather long life times of the $4^{14}O$ and ^{15}O oxygen isotopes. This causes enrichment in ^{14}O and ^{15}O which is observed as nitrogen enrichment in the ejected material. The actual energy generation rate is limited by the β-decay rates but is also affected by the hydrogen and CNO fuel available. A more quantitative interpretation of the actual nucleosynthesis therefore requires detailed knowledge of the associated convective processes which may transport freshly produced material out of the actual hot burning zone or may bring more hydrogen fuel into the burning zone. In addition, the fuel balance may also change due to additional proton capture processes on short-lived radioactive nuclei in the CNO range. Of particular relevance are the time scales for reaction sequences like $^{16}O(p,\gamma)^{17}F(p,\gamma)^{18}Ne(\beta,\nu)^{18}F(p,\alpha)^{15}O$ which would control fast additional fuel supply for the hot CNO cycle. Observations of over-abundances in the Ne to S mass range characterize Ne-novae which are interpreted to

---

[3]The Eddington luminosity, also referred to as the Eddington limit, is the maximum luminosity a body (such as a star) can achieve when there is balance between the force of radiation acting outward and the gravitational force acting inward. The state of balance is called hydrostatic equilibrium. When a star exceeds the Eddington luminosity, it will initiate a very intense radiation-driven stellar wind from its outer layers. Since most massive stars have luminosities far below the Eddington luminosity, their winds are mostly driven by the less intense line absorption. The Eddington limit is invoked to explain the observed luminosity of accreting black holes such as quasars.

[4]S. Shore et al.: Astrophys. J. 421, 344 (1994)

be thermonuclear runaways on accreting O-Ne white dwarfs with infusion of oxygen and neon (and magnesium) into the accreting envelope. Proton capture reactions on the initial ^{20}Ne and ^{24}Mg lead to the production of heavier isotopes up to ^{32}S. This agrees well with recent observations in nova ejecta of silicon and sulfur. According to theoretical model simulations considerable production of the long-lived radioisotopes ^{22}Na and ^{26}Al is expected in Ne novae. However, observations of the gamma activity in novae with the COMPTEL observatory gave no indication for ^{22}Na or ^{26}Al activity. There is an order of magnitude discrepancy between predicted and observed intensities of the ^{22}Na gamma ray[5]. Studies of the nuclear reactions crucial to the synthesis of these radioisotopes are therefore particularly important. Self-consistent calculations for nova nucleosynthesis nuclear reaction rates clearly indicate significant impact of nuclear reaction rates on fast nova nucleosynthesis. Current nova models predict significantly less mass of ejected material than is observed. Solutions to this problem suggest that peak temperatures in nova explosions may be much higher than the accepted 400 million degrees K. Such high temperatures imply that break-out of the hot CNO cycle can occur for some novae - agreeing with at least one recent nova observation. The nuclear physics information required to understand this higher-temperature burning includes measurements of nuclear reactions on proton-rich unstable isotopes. These reaction rates still carry considerable uncertainties since they are mostly estimated on the basis of nuclear structure and nuclear level analysis[6].Study of the impact of nuclear reaction rates on nova nucleosynthesis indicated significant uncertainties for abundance predictions of all isotopes above mass A=20. The present reaction rates are reliable only for predictions of Li, Be, C, and N abundances. Many of the reaction rate uncertainties have to be reduced in order to predict more reliable O, F, Ne, Na, Mg, Al, Si, S, Cl, and Ar abundances.

8.3   X-Ray Bursts

X-ray bursts are explained as thermonuclear runaways in the hydrogen rich envelope of an accreting neutron star. Figure 8.1 shows the observed light curve of a single X-ray burst which rapidly increases within seconds to maximum intensity. Low accretion rates favor a sudden local ignition of the ma-

---

[5]N. Prantzos, R. Diehl: Phys. Rep. 267, 1 (1996)

[6]C. Iliadis, J.M. D'Auria, S. Starrfield, W.J. Thompson, M. Wiescher: Astro- phys. J. Suppl. 134, 151 (2001)

terial with a subsequent rapid spread over the neutron star surface[7]. This is also indicated in Figure 8.1 which shows the spreading of the burning front over the entire neutron star surface.

8.4 Thermonuclear Runaway.

The thermonuclear runaway is triggered by the ignition of the triple-alpha reaction and the break-out reactions from the hot CNO cycles. Therefore the on-set of the X-ray burst critically depends on the rates of the alpha capture reactions on ^{15}O and ^{18}Ne. The rates for these break-out reactions carry still considerable uncertainties [251,127] and are subject to intense experimental studies using a wide variety of techniques with stable and radioactive beams. The $^{15}O(\alpha,\gamma)^{19}Ne(p,\gamma)^{20}Na$ reaction sequence represents the main link by which the initial CNO isotopes will be converted towards heavier elements. With the onset of the $^{14}O(\alpha,p)^{17}F$ reaction a continuous flow of ^{4}He towards ^{21}Na via the sequence $^4He(2\alpha,\gamma)^{12}$ $C(p,\gamma)^{13}N(p,\gamma)^{14}O$ $(\alpha,p)^{17}F(p,\gamma)^{18}Ne(\alpha,p)^{21}Na$ by-passes the $^{15}O(\alpha,\gamma)$ link and $^{18}Ne(\alpha,p)^{21}Na$ emerges as the main break-out reaction controlling the flow towards heavier masses. The uncertainty of the $^{15}O(\alpha,\gamma)^{19}Ne$ rate depends on the contribution of a single low energy resonance at 4.033 MeV excitation energy which remains elusive to experimental scrutiny. Most of the estimates are based on the α strength of the mirror state in ^{19}F which was determined from $^{15}N(^6Li,d)^{19}F$ transfer measurements. The uncertainty in the $^{18}Ne(\alpha,p)^{21}Na$ rate depends on the contributions of a fairly large number of low energy resonances which have not been studied yet directly. In particular no clear spin and parity assignment has been made yet. This makes a reliable estimate of the reaction rate rather futile since only natural parity states will be contributing to the $^{18}Ne + \alpha$ reaction channel. The thermonuclear runaway itself is driven by the αp-process and the rapid proton-process (short rp-process) which convert the initial material rapidly to ^{56}Ni causing the formation of Ni oceans at the neutron star surface. The α p-process is characterized by a sequence of (α,p) and (p,γ) reactions processing the ashes of the hot CNO cycles ^{14}O and ^{18}Ne up to ^{34}Ar and ^{38}Ca range. Except for $^{14}O(\alpha,p)^{17}F$ and $^{18}Ne(\alpha,p)^{21}Na$ the reaction rates are all based on Hauser Feshbach predictions. The validity of the statistical approach in this mass range has to be tested. Since pronounced α-cluster structure in the T=1 even-even nuclei near the α threshold may lead to the occurrence of pronounced low energy resonances, waiting points at positions

---

[7]L. Bildsten, T. Strohmayer: Physics Today 52, 40 (1999)

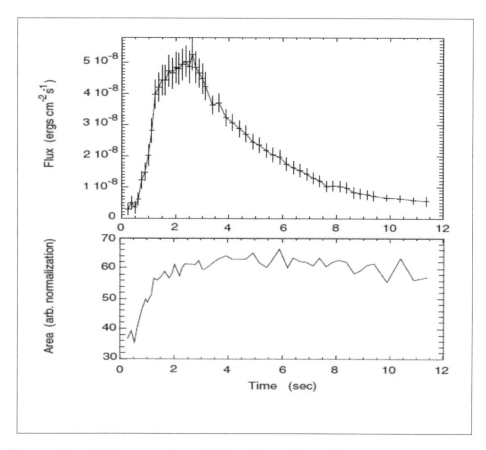

Figure 8.1: A thermonuclear X-Ray Burst from the neutron star in the low mass xray binary system 4U 1728-34, as observed with the Rossi X-ray Timing Explorer. The top shows the rapid increase of the x-ray flux, followed by a slower decay. The lower panel indicates the change of the x-ray emitting area calculated for blackbody radiation from a special system. The initial increase of the area provides strong evidence of the spread of the nuclear burning front over the entire surface of the neutron star. Ref: Physics Today 52, 40 (1999)

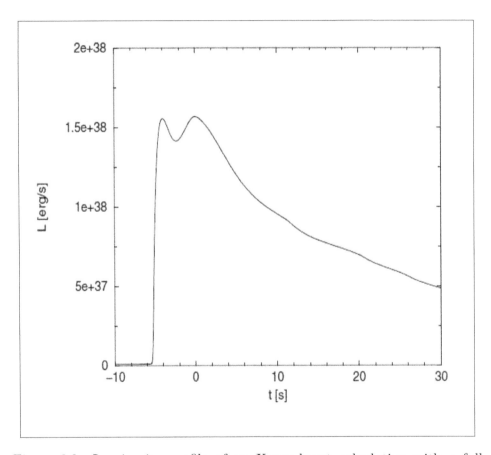

Figure 8.2: Luminosity profile of an X-ray burst calculation with a full nuclear network, assuming a neutron star of $1.4 M_\odot$ and a mass accretion rate of. The double peak structure (often seen in observations) is here due to the waiting point ^{30}S. Variations in neutron star size and accretion rate can change this structure, producing a regular single burst peak as well.

where nuclear reaction flows are halted for beta-decay lifetimes, because (p, γ)- and α-induced reactions are either not competitive or overbalanced by a faster reverse reaction, have the potential to be observed in the detailed peak structure of X-ray light curves as shown in Figure 8.2.

8.5   The rp-Process.

The rp-process represents a sequence of rapid proton captures up to the proton drip line and subsequent β-decays of drip line nuclei processing the material from the argon, calcium range up to ^{56}Ni and beyond (see Figure 5.5). The flow is halted at ^{56}Ni during peak temperatures of around 2.0 to 3.0 billion degrees because of a $(p, \gamma) - (\gamma, p)$ chemical equilibrium due to a small reaction Q-value. In the subsequent cooling phase of the explosion photo disintegrations are suppressed and the rp-process continues beyond ^{56}Ni. The nucleosynthesis in the cooling phase of the burst alters considerably the abundance distribution in atmosphere, ocean, and subsequently crust of the neutron star. This may have a significant impact on the thermal structure of the neutron star surface and on the evolution of oscillations in the oceans. To verify the present models nuclear reaction and structure studies on the neutron deficient side of the line of stability are essential. Measurements of the break-out reactions will set stringent limits on the ignition conditions for the thermonuclear runaway, measurements of alpha and proton capture on neutron deficient radioactive nuclei below ^{56}Ni will set limits on the time-scale for the actual runaway but will also affect other macroscopic observables. Simulations of the X-ray burst characteristics with self consistent multi-zone models suggest indeed a significant impact of proton capture reaction rates between A=20 and A=64 on expansion velocity, temperature and luminosity of the burst. The presently suggested reaction rates carry enormous uncertainties. Shell model based calculations of proton capture rates in this mass range[8] did indeed indicate up to several orders of magnitude discrepancies to the global Hauser Feshbach predictions[9]. Hence, more experimental data are necessary to remove the uncertainties. Nuclear structure and nuclear reaction measurements near the double closed shell nucleus ^{56}Ni determine the conditions for the flow to heavier nuclei in the cooling phase. This depends on the actual rate of $^{56}Ni(p, \gamma)$ and the Q-value which determine the details of the chemical or $(p, \gamma) - (\gamma, p)$ equilibrium. First measurements with ^{56}Ni beams lead to improved estimates of

---

[8]Phys. Rev. C 52, 1078 (1995)
[9]Phys. Rep. 294, 168 (1998)

the reaction rate but considerable uncertainty still remains. Structure and reaction measurements beyond ^{56}Ni, in particular the experimental study of 2-proton capture reactions bridging the drip-line for even-even N = Z nuclei like $^{64}Ge, ^{68}Se$ and ^{72}Kr are necessary to determine the final fate of the neutron star crust. The 2p-capture rates depend sensitively on the proton binding energies of the $^{65}As, ^{69}Br$ and ^{73}Rb nuclei, therefore reliable mass measurements are necessary for nuclei along the proton drip line in this mass range. These mass measurements have to be complemented with decay studies. Of particular importance are beta-decay studies of isomeric and/or thermally populated excited states, which are not accessible for experiment with available equipment. In particular capture reactions on isomeric states may cause a significant change in reaction flow. There is a substantial need for nuclear structure information at the proton drip line, especially in the mass region along the drip line up to the A=100 range, to address the multitude of open questions on the nucleosynthesis path and pattern in the rp-process towards its endpoint. An important question is the one for the endpoint of the rp-process. The endpoint is determined both by the macroscopic time-scale of the burst which depends on the various cooling mechanisms and the microscopic time-scale given by the effective life-times of the waiting point nuclei along the reaction path. For steady state burning scenarios, characterized by long-term high temperature conditions, the endpoint is basically determined by the availability of hydrogen fuel in the accreted layer. All of the predicted reaction and decay rates along the reaction path need to be experimentally tested and verified, yet, the present predictions suggest a reaction flow even beyond ^{100}Sn. This raises the question for the actual endpoint. Alpha decay studies in the mass range above ^{100}Sn suggest that the neutron deficient isotopes ^{106}Te to ^{108}Te and ^{108}I predominantly decay into the α channel. The isotope ^{109}I has been identified as a short-lived proton emitter. These observations suggest that the actual endpoint of the rp-process might be associated with rapid back-processing of the material via proton or γ induced α decay in the Sn, Sb, Te range near the proton drip line. The most likely processes are $^{104}Sb(p,\alpha)^{101}Sn$ or $^{105}Te(\gamma,\alpha)^{101}Sn$, $^{105}Sb(p,\alpha)^{102}Sn$ or $^{106}Te(\gamma,\alpha)^{102}Sn$, and possibly also $^{106}Sb(p,\alpha)^{103}Sn$. While some mass models predict ^{104}Sb and ^{105}Sb to be proton unbound, the experimental decay-rates are longer than Hauser Feshbach predictions for the (p,α) reaction rates. Only a small amount of ^{106}Te, ^{107}Te is produced by $^{105}Sb(p,\gamma)$ and $^{106}Sb(p,\gamma)$, respectively and will be rapidly depleted by $^{106}Te(p,\alpha)^{103}Sb$, and $^{107}Te(p,\alpha)^{104}Sb$. Detailed reaction flow calculations for the rp-process have been performed to simulate the reaction flow in this mass range. It was clearly shown that multi-cycling emerges in the Sn-Te-I range. This cycle presents a strong impedance for the rp-process and has been identified as the actual end-point for the reaction path.

8.6 X-Ray Pulsars

X-ray pulsars are interpreted as accreting neutron stars with high accretion rates and large magnetic fields which funnel the accreted material to the pole cap, thus reaching locally high accretion rates. This causes a steady burning of the accreted material via the αp- and the rp-processes on the surface of the neutron star. Detailed studies of the nucleosynthesis suggest that the accreted material is rapidly converted to heavier elements in the mass 80 to 100 range which changes drastically the composition of the crust and the ocean of the neutron star replacing the original iron crust by a mixture of significantly more massive elements. Thus the composition of the neutron star crust in a binary system is substantially different from the one in a primordial single neutron star. This may have important effects on the thermal and electromagnetic conditions at the neutron star surface. The modified mass composition may also affect the observed decay of the magnetic field of the neutron star and its rotational r-modes due to shear effects. The final composition depends strongly on the nuclear physics associated with the rp-process and the endpoint of the rp-process, which is directly correlated with the accretion rate. For experimental confirmation in the lower mass range, studies similar to those for the X-ray burst simulations are required. However, for accretion in excess of fifty times the Eddington limit, the endpoint of the rp-process is expected to lie in the mass 150 range. This requires a new range of nuclear structure data near the limits of stability. Of particular interest are beta-decay lifetimes, and especially processes like direct proton and alpha decays. If these processes dominate, as is expected for the decay of very neutron deficient Tellurium, Iodine, and Xenon isotopes a natural halting point for the rp-process is reached.

8.7 Accretion Processes on Neutron Stars and Black holes

Many of the accretion processes on neutron stars and black holes take place through accretion disks. Therefore it is necessary to investigate nuclear interaction processes that may occur during the accretion process at the local accretion disk in low density and high temperature conditions. Such processes may be of relevance for modeling X-ray bursts and X-ray pulsar events since such processes may alter the abundance distribution of the accreted material substantially[10]. At the fairly high temperature conditions of an accretion disk a considerable amount of the accreted hydrogen may actually be consumed before entering the neutron star atmosphere. This may affect our current understanding of the nucleosynthesis and energy release following ignition of the accreted material in thermonuclear runaways or steady burning modes. On the other side interaction of the high velocity accreted material with the outer atmospheres of the accreting object may cause spallation and fragmentation of the accreted material which in turn may greatly alter its composition and regenerate part of the pre-consumed hydrogen. Clearly detailed nucleosynthesis studies in the accretion disk and during impact on the outer atmosphere of a neutron star are necessary to address these problems.

References

Quoted in text:

- J.W. Truran: In: Essays in Nuclear Astrophysics, eds. C.A. Barnes, D.D. Clayton, D.N. Schramm (Cambridge University Press, Cambridge 1982) p. 467

- S. Shore et al.: Astrophys. J. 421, 344 (1994)

- S. Starrfield, J.W. Truran, M. Wiescher, W.M. Sparks: Mon. Not. R. Astron. Soc. 296, 502 (1998)

- R. Rosner, A. Alexakis, Y.-N. Young, J.W. Truran, W. Hillebrandt: Astrophys.J. 562, L177 (2001)

- S. Starrfield, C. Iliadis, J.W. Truran, M. Wiescher, W.M. Sparks: Nucl. Phys. A 688, 110c (2001)

- N. Prantzos, R. Diehl: Phys. Rep. 267, 1 (1996)

- Mike Guidry, Stellar Structure and Early Stellar Evolution, 2008, unpublished manuscript, Department of Physics and Astronomy, University of TennesseeKnoxville

- A.E. Champagne and M. Wiescher, Annual Review of Nuclear and Particle Science, 42:39-76, 1992 101

---

[10]R.E. Taam, B.A. Fryxell: Astrophys. J. 294, 303 (1985)

Chapter 9

Formation of stars and galaxies

9.1 Introduction

To the science communities, the beginning of the 21st century is really a great time for many new ideas, experiments, theories, measurements and observations are being converged to what appears to be a composite picture of the Universe we duel in. During the last 4-5 decades we were able to e accumulate more scientific knowledge than throughout the rest of our lifetime on this planet. One of the greatest outcomes of this process is that we are in a position to explain the evolution of the universe from the time when it was a mere 10^{-43} second old baby and all the way up to the emergence of intelligent life forms. Since the answers obtained have always been under scanner for posing more and more intelligent questions about the outcomes, constant scientific efforts, resulted a remarkably insightful understanding of the laws of Nature. Our curiosity, thirst and zeal for knowledge have gifted us a modern version of Genesis. Our theories describe how the Big Bang started a Universe that would later form clusters, galaxies and apparently countless quantities stars. The evolution of these stars, process the light elements hydrogen and helium into the elements we observe today, including the elements needed for the development of life. This modern Genesis directly or indirectly teaches us that we all are offspring of the stars. One of the hot topics in nuclear astrophysics is the synthesis of elements which can be better understood by the blend of large-scale hydro dynamical simulations, nuclear theories and experiments, knowledge of atomic transition lines, and an availability of very accurate observational data.

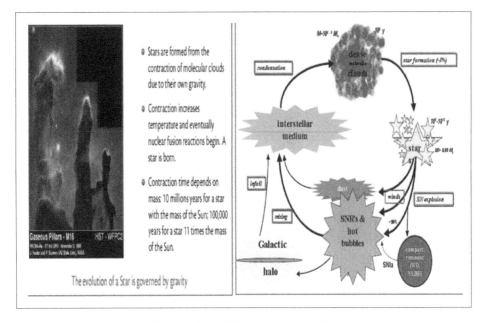

Figure 9.1: The cosmic cycle.

9.2   Life cycle of Stars

A star is a luminous sphere of plasma held together by its own gravity. The nearest star to Earth is the Sun. Other stars are visible to the naked eye from Earth during the night, appearing as a multitude of fixed luminous points in the sky due to their immense distance from Earth. Nebulae within galaxies are gigantic clouds of gas and dust. Gravitation or some external event can cause part of these clouds to contract. The mass of gas concentrates and the molecules collide; the temperature rises until hydrogen fusion occurs. Star birth occurs when giant gas clouds start to clump together and gravity begins to pull the clumps toward each other. This clumping is often caused by energy waves from nearby exploding supernovae. These clouds are made of mostly hydrogen and some helium. These are the fuel a star uses to shine brightly. Astrophysicists can see the birth of stars because of the infrared radiation emitted by the stars through these clouds. Stars form when giant gas clouds start to clump together and gravity begins to pull the clumps toward each other. This clumping is often caused by energy waves from nearby exploding supernovae. These clouds are made of mostly hydrogen and some helium. These are the fuel a star uses to shine brightly. Galaxies are system of millions or billions of stars, together with gas and dust, held together by gravitational attraction. Since, formation of galaxy is a core theme of cosmology (See Figure 9.1), they provide the beacon(lighthouse) with which we can measure the expansion and acceleration of the universe. The present understanding regarding the formation of our solar system is that, a nearby supernova triggered the condensation of our Sun and the planets from a giant molecular cloud (GMC), and provided some of the heavy elements we are familiar with. This is supported by the fact that the decay product of iron-60, nickel-60, has been found on Earth. Iron-60, which has a half-life of 2.6 million years, could have been created only under supernova conditions. Stars, the most fundamental objects of the Cosmos, are also referred as homes of the planetary systems. They provide the energy necessary for the development and sustenance of life. The evolution of stars drives the evolution of all stellar systems including clusters and galaxies. Stellar Evolution also controls the chemical evolution of the universe. The physical natures and life histories of stars are well described by the theory of stellar structure and evolution, one of the great achievements of 20th century astronomy. According to this theory stars are self-gravitating balls of (mostly) hydrogen gas that act as thermonuclear furnaces to convert the primary product of the big bang (hydrogen) into the heavier elements of the periodic table. For most of its life a star maintains a state of stable equilibrium in which the inward force of gravity is balanced by the outward force of pressure. This internal pressure is generated by the energy released in the nuclear reactions which burn hydrogen at the star's center. These nuclear reactions are also the source of the star's luminosity (i.e., energy output per second). During the time the star burns hydrogen in its core it maintains a fixed radius and luminosity and consequently a constant surface temperature. The exact value of the equilibrium radius, luminosity and surface temperature of a star depends almost exclusively on one parameter, the star's mass. The greater the mass of a star, greater will be its luminosity, size and surface temperature. In a group, hydrogen burning stars of varying mass form a well defined locus of points in the observable luminosity-effective temperature plane called the HR diagram (see Figure 9.2). The locus of points is called the main sequence and the hydrogen burning phase of a star's life is known as the main sequence phase. Main sequence stars range in mass from about 0.08 to $100 M_{\odot}$. Stars with smaller masses have insufficient weight to raise their central temperatures enough to enable hydrogen fusion are referred as brown dwarfs. Stars with larger masses are presumably too luminous to hold on to their outer atmospheres. Once a star exhausts its supply of hydrogen in its central core, nuclear reactions there crash down (falls off) and the helium-rich core contracts became unable to support itself against gravity.

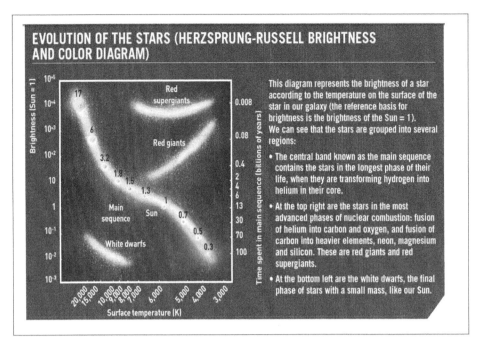

Figure 9.2: The Herzsprung- Russel diagram. [Source: www.google.com]

The contraction of the core releases large amounts of gravitational potential energy that cause an increase in hydrogen burning in a shell surrounding the core. For star like our sun, this eventually leads to the expansion of the star's outer layers which is accompanied by a drastic increase in the star's luminosity (See Figure 9.3). This begins the post-main sequence phase of stellar evolution during which the star evolves away from the main sequence while keeping its temperature roughly constant in a fashion similar to that Hayashi et al.[1]. The inner core gains mass from the continual production of helium in the hydrogen burning shell and contracts until fusion reactions involving helium restore a temporary equilibrium in which gravity is again balanced by internal pressure. After this phase, such stars eject their outer atmospheres producing planetary nebula and leaving behind a white dwarf stellar remnant. More massive stars experience a more complex post-main sequence evolution during which they will fuse heavier elements in their cores and eventually end their lives in catastrophic supernova explosions which violently eject much of their original mass, enriched by heavy elements, into

---

[1]Hayashi, C. and Hoshi, R.: PASJ, 13, 442(1961) NB:PASJ stands for Publications of the Astronomical Society of Japan

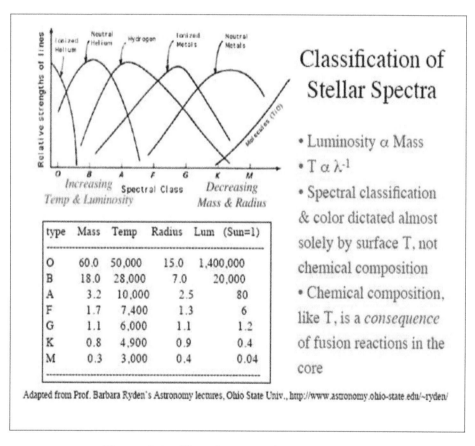

Figure 9.3: Classification of stellar spectra.

the interstellar medium leaving behind exotic remnants, namely neutron
stars and black holes. Despite its spectacular success in explaining the life
histories and deaths of stars, the theory of stellar evolution is incomplete in
a very fundamental aspect. It is not able to account for the origin of stars.
Knowledge of the physical mechanism for the formation of stars is essential
for understanding the evolution of the galaxies and the universe from the
earliest times after the big bang to the current epoch of cosmic history. Development of a theory of star formation is also crucial for understanding the
origin of planetary systems which, in turn, is important for evaluating the
possibility of biology beyond the solar system. The inability of the theory
of stellar evolution to explain star formation, likely points to the inherent
complexity of the physical process itself. Consequently construction of a
theory of star formation must require a strong foundation of empirical data
or observation. The empirical study of star formation is greatly facilitated
by a fundamental property of the universe. Namely that star formation has
been a continuous and ongoing process which in our galaxy extends into the
present epoch. Consequently, the physical process of star formation can be
investigated by direct observation. The realization of ongoing star formation
in our galaxy was an important milestone in twentieth century astronomy
and as such merits some further discussion. By the middle of the twentieth
century the theory of stellar structure and evolution had demonstrated that
certain luminous stars, OB stars (as classified by an Armenian astronomer
V. A. Ambartsumian in 1947) , burn their nuclear fuel at such prodigious
rates that they can live for only a small fraction of the lifetime of our galaxy.
The very existence of such stars clearly indicated that star formation has
occurred in the present epoch of Galactic history. The space densities of
stars in OB associations were well below the threshold necessary to prevent
their disruption by Galactic tidal forces. The calculated dynamical ages for
the OB associate stars were found to be much less than the age of the galaxy.
These dynamical ages turned out to be in good agreement with the nuclear
ages of the stars and independently provided evidence that star formation
is still an active process in the Galaxy. The discovery of the interstellar
medium of gas and dust during the beginning of 20th century provided a
crucial piece of corroborating evidence in support of the concept of present
epoch Galactic star formation. Subsequent observations of interstellar material established that clouds of interstellar gas and dust had roughly stellar
composition and were considerably more massive than a single star or group
of stars. This observation reveals the fact that the raw materials required to
manufacture new stars was relatively abundant in the Galaxy. The stellar
evolution theory, expanding OB associations and the interstellar medium,

constitute basic proofs of ongoing star formation in our Milky Way galaxy. This concept has also been advanced by Walker[2] who noticed a large population of low mass stars lying above the main sequence in the HR diagram of the young cluster NGC2264, consistent with the predictions of Henyey et al.[3]] and later Hayashi[4] who showed that the locations of these stars in the HR diagram indicated that they were contracting pre-main sequence stars that had not yet initiated fusion reactions in their cores. For much of the last 50 years direct observation of the star formation process and the development of a theory to explain it, have been severely hampered by the fact that most stars form in dark clouds and during their formative stages they are optically invisible. Fortunately, advances in observational technology over the last quarter century opened the infrared and millimeter-wave windows to astronomical investigation and enabled direct observations of star forming regions and this has significantly expanded our knowledge of the star formation process. As a result the foundations for a coherent theory of star formation and early evolution are being laid as demonstrated in Figure 9.4. Stars are formed from the contraction of molecular clouds due to their own gravity. Contraction increases temperature and eventually nuclear fusion reactions begin, a star is born. Contraction time which depends on the mass of the star has a typical value of about 10 millions years for a star with the mass of the Sun and 100,000 years for a star 11 times the mass of the Sun. The evolution of a star is governed by gravity. When observed at a dark site on a clear moonless night the Milky Way is truly an impressive sight. One its most prominent characteristics is that it is split down the middle by a dark obscuring band. The band consists of the superposition of many interstellar dark clouds which contain tiny opaque dust grains that very effectively absorb and scatter the background starlight. These dark clouds are the sites of star and planet formation in the galaxy. A survey revealed three fundamental results pertaining to star formation which states that

- i) Stars form in dense $(n(H_2) \geq 10^4 cm^{-3})$ molecular gas,

- ii) Most stars form in embedded clusters, and

- iii) The overall star formation efficiency in GMCs is quite low, of order 1-3% or less.

---

[2]Walker, M. F.: ApJS, 2, 365(1956)

[3]Henyey, L. G., Lelevier, R., and Levee, R. D.:PASP, 67,154 (1955)

[4]Hayashi, C.: PASJ, 13, 450(1961)

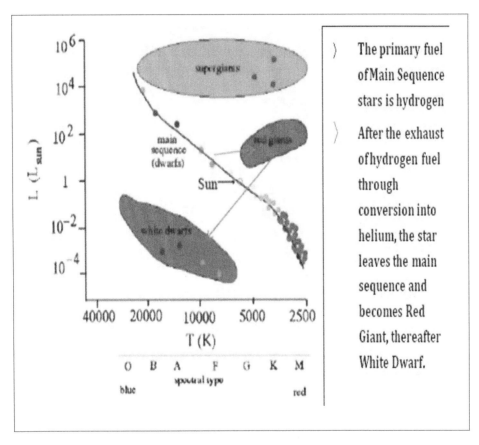

Figure 9.4: Stellar evolution.

Salient points to remmeber

- The chemical elements are sprayed into space by three or more processes, each involving a dying star:

  1. Main sequence stars like our Sun burn out and become cold white dwarves. But before their death, they pass through a red giant stage where the outer mantle layers, enriched in elements like oxygen, nitrogen and carbon – often in the form of nanometer size diamonds – are quietly 'blown off' into the interstellar medium.

  2. Stars eight times heavier than our Sun explode as a super nova. Due to the thermo chemistry of the various nuclear processes, each shell of nucleosynthesis proceeds on an accelerating time scale and Si burns to Fe in few hours. Conditions in the core become so extreme that electron pressure is overcome and the protons are forced to react with electrons to produce neutrons $p+e \rightarrow n+\nu_e$ and a neutron star is born in less than a second. The rebounding shock wave plus radiation pressure from the escaping neutrinos causes the outer layer stars to explode outwards as a Type II supernova. These conditions have a massive flux of free neutrons and the various nuclei are able absorb one or more of these neutrons, undergo beta decay, absorb another neutron or neutrons, another beta decay. A process which moves nuclei up (heavier) the periodic table towards and past uranium.

  3. The third process is the Type Ia supernova. This occurs when a white dwarf is held in a tight binary association with a main sequence star. The small, dense white dwarf pulls the surface layers from the companion star until enough mass builds so that a runaway thermonuclear incineration occurs on the surface of the white dwarf which explosively disassembles.

- During a supernova, the process of neutron capture builds up elements as far as Z=100, Ferminium. The nucleus ^{257}Fm has a half-life of a few months, but apparently the process of nucleosynthesis cannot go no further.

- In each case, shells of debris - consisting of a isotopic zoo of atomic nuclei – are ejected into the interstellar medium. Over millions of years this material cools to a mildly radioactive solid ashes that gathers together by gravity... where it participates in next generation of star formation.

- For reasons not yet fully understood, the contracting cloud of hydrogen, helium and metals - astronomers regard all elements other then H and He as metals - evolves to form a disk of planets around a central star.

- The inner planets of our solar system are not massive enough to have sufficient gravity to hold onto hydrogen and helium, and these gases escape into space leaving the Earth depleted in these elements, but enriched in heavy elements with respect to the universe as a whole. Our planet is large enough so that the residual radioactivity, mainly from ^{40}K, is able to heat the mantle so that it remains hot and fluid.

- Some hydrogen remains on Earth by chemically combining with oxygen and trapped as water, a substance essential for our planet flora and fauna. When the Earth formed much of the hydrogen was combined with carbon as methane, CH_4, however this is a comparatively rare substance today.

Solved Problems

1. In a rough but reasonable, approximation, a neutron star may be considered as a sphere which consists almost entirely of neutrons which form a nonrelativistic degenerate Fermi gas. The pressure of the Fermi gas is counterbalanced by gravitational attraction.

   (a) Estimate the radius of such a star to within an order of magnitude if the mass is 10^{33} g. Since only a rough numerical estimate is required, you need to make only reasonable simplifying assumptions like taking a uniform density, and estimate integrals you cannot easily evaluate, etc.

   (b) In the laboratory, neutrons are unstable, decaying according to $n \rightarrow p + e^- + \overline{\nu}_e + 1 MeV$ with a lifetime of 10^{-10} s. Explain briefly and qualitatively, but precisely, why we can consider the neutron star to be made up almost entirely of neutrons, rather than neutrons, protons, and electrons.

   Solution (a) Let R be the radius of the neutron star which is assumed spherical. The gravitational potential energy is

$$V_g(R) = -\int_0^R (G/r)(\frac{4\pi r^3 \rho}{3})(4\pi r^2 \rho dr) = -(\frac{3}{5})(\frac{GM^2}{R}),$$

   where $\rho(= \frac{3M}{4\pi R^3})$ is the density of the gas, M being its total mass, G is the gravitational constant. When R increases by dR, the pressure P of the gas performs an external work $dW = PdV = P4\pi R^2 dR$. Again since, $dW = -dV_g = P4\pi R^2 dR$. We have

$$P = -\frac{1}{4\pi R^2}\frac{dV_g}{dR} = \frac{3GM^2}{20\pi R^4}.$$

The pressure of a completely degenerate Fermi gas is

$$P = \frac{2}{5}NE_F,$$

where $N = \rho/M_n$ is the number density of neutron, M_n being the neutron mass,

$$E_F = \frac{\hbar^2}{2M_n}\left(\frac{9\pi}{4}\frac{M}{M_n R^3}\right)^{2/3}$$

is the limiting energy. Equating the expressions for P we get

$$
\begin{aligned}
R &= \left(\frac{9\pi}{4}\right)^{2/3}\left(\frac{\hbar^2}{GM_n^3}\right)\left(\frac{M_n}{M}\right)^{1/3} \\
&= \left(\frac{9\pi}{4}\right)^{2/3} \times \frac{(1.05 \times 10^{-34})^2}{6.67 \times 10^{-11})^3} \times \left(\frac{1.67 \times 10^{-27}}{10^{30}}\right)^{1/3} \\
&= 1.6 \times 10^4 m
\end{aligned}
$$

(b) (b) Let d be the distance between neighboring neutrons. As $\frac{M}{M_n} \simeq (2R/d)^3$, $d \simeq 2R(M_n/M)^{1/3} = 4 \times 10^{-15}m$. If electrons existed in the star, the magnitude of their mean free path would be of the order of d, and so the order of magnitude of the kinetic energy of an electron would be $E \simeq pc \sim c/d \sim 50MeV$. Since each neutron decay only gives out 1 MeV, and the neutron's kinetic energy is less than $E_F \simeq 21$ MeV, it is unlikely that there could be electrons in the neutron star originating from the decay of neutrons, if energy conservation is to hold. Furthermore, because the neutrons are so close together, e and p produced from a decay would immediately recombine. Thus there would be no protons in the star as well.

Exercise

1. Solve above problem (1) for another neutron stars of mass i) 10 times, ii) 100 times, iii) 200 times, iv) 500 times, and v) 1000 times more massive than the above and cooment on the result. Graphically show how the radius depends on the mass of the star.

2. Derive an expression for the estimation of the luminosity of a star.

3. Explain the phyisical significance of the HR diadram. What is meant by 'Chandrasekhar mass'? Under what conditions a star's life ends to-i) white dwarf, ii) red giant, iii) black hole?

References

Quoted in the text:

- Hayashi, C., & Hoshi, R. 1961, PASJ, 13, 442

- Walker, M. F. 1956, ApJS, 2, 365

- Henyey, L. G., Lelevier, R., & Levee, R. D. 1955, PASP, 67, 154

- Hayashi, C. 1961, PASJ, 13, 450 **Further Readings:**

- Introduction to Modern Astrophysics by Carroll & Ostlie, Addison-Wesley (1996)

Chapter 10

Role of Computation in Nuclear Astrophysics

10.1 Introduction

The dramatic impact of computation on astronomy and astrophysics is manifested in many ways. Modern numerical codes are now being used to simulate and understand the evolution, explosion, and nucleosynthesis of stars, how the elements are injected into the interstellar medium, molecular clouds, and extant planetary systems, and the cosmic evolution of the abundances. They are also essential to processing astronomical spectral databases whose sizes now exceed one terabyte into abundance data that are usable by the nuclear astrophysics community. The largest codes may have in excess of a million lines and run on supercomputers that have more than 1,00,000 cores, generating datasets that occupy one to a hundred terabytes of storage. The most widespread codes may be driven by communities with hundreds of users and run on desktop class machines that generate a significant fraction of the published literature. Such codes are now an indispensable part of the nuclear astrophysics enterprise. They often deploy teams of astronomers, astrophysicists, computer scientists, visualization professionals, applied mathematicians, and algorithm specialists to create, maintain, and constantly develop them.

NSF, NASA, and DOE have made substantial investments in the advanced computing and networking ecosystem over the last few decades, from national to regional to university to individual facilities. Sustained petascale, and soon exa-scale, computing capabilities will be available to the nuclear astrophysics community. Such capabilities will enable cutting-edge theoretical calculations and analyses that push the nuclear astrophysics frontier. One example is 3D core-collapse simulations to better quantify the neutron star winds, the r-process signatures, and evolution of the proto-neutron star. Another example is routinely deploying reactions networks with 1000's of isotopes in all 2D or 3D models. Future progress in advanced computing for nuclear astrophysics will come from further parallelization, ubiquitous deployment of next-generation 100 GB/s internet connectivity in tandem with Globus Online, distributed cloud storage systems, extracting actionable knowledge from big data, and, potentially, social computing. Similarly, Spectroscopic stellar surveys that are scheduled over the next 5-10 years, or are recently completed (SEGUE, LAMOST, APOGEE, HERMES, GAIA, a variety of LSST follow ups, and the proposed 10 meter spectrographic survey telescopes), as well as all-sky surveys that will provide significant new information on supernovae and other transient events (LSST, Sky Mapper). Nuclear astrophysics will have terabytes of spectroscopic data and petabytes of photometric all-sky data that need to be analyzed and compared to simulations.

These new technological capabilities and data driven science will enable qualitatively new physical modeling in topics relevant to nuclear astrophysics. Exploiting these new capabilities for nuclear astrophysics will require new software instruments (e.g., run-time visualization for 100TB of 1PB data sets) and sustained funding support for focused multi-institutional research collaborations. The rapid advances in computers are the key to developments of the complex theories of nuclear structure, supernova explosions, and cosmic evolution of the abundance of the elements: super fast calculations enable theories and their parameters to be widely explored, and to guide experiments and compare their results with theoretical predictions. Also support for databases of thousands of nuclei and their reactions, and advanced graphics-processing to visualize complex data or theories, are key drivers of the field. The emerging computer technology and numerical methods have fundamentally changed the way certain investigations are carried out in physics. Instead of adhering to models which are solvable analytically, investigators can now easily explore new ideas that require extensive calculations and large-scale simulations. As example we may refer to the lattice gage studies for transition from hadronic matter to quark-gluon plasma (QGP). In fact, numerical solution using the Feynman path integral approach is likely to be the only way that answers can be obtained for some of the strong interaction problems.

Theoretical investigations of eigenvalue problems in nuclear physics often involve intensive computing because of their highly nonlinear nature.

Partly, it is also because of the non perturbative nature of the phenomena involving strong inter-particle interactions. Here, computing is actually an advantage to carry out large calculations which can be delegated to machines that are many a degree faster and more reliable than human being. This applies to numerical computations as well algebraic manipulations. In fact, many of the algebraic calculations involved in analytical solutions are often carried on computers as well. Furthermore, if we take good advantage of the visualization tools available on computers, many complicated solutions, both analytical and numerical, can be more readily comprehended than with algebraic symbols alone. In many cases, far more advantage of the visualization capabilities can be used in understanding the results from complicated calculations than what have been done in practice. For example, to be able to "see" a multi-dimensional result is certainly not something that can be achieved easily without computers. In addition to lattice gauge calculations, several problems in nuclear theory can also benefit from intensive computation. The relativistic shell model is one such example. In the study of nuclear structure, physicists mostly deal with velocities that are much less than the speed of light. The relativistic effect can nevertheless be important in such cases from the following considerations. In the Schrodinger picture, each spin-1/2 fermion has two components, one with spin pointing up and the other with spin pointing down. In the more general Dirac picture, the corresponding wave function has four components, two for the particle and two for the antiparticle. The four-component Dirac equation may be expressed as a set of two coupled equations, each one having only two components. At non-relativistic energies, the coupling between the "upper" two components and the "lower" two components may be replaced by a spin-orbit term and the two equations decouple from each other in this approximation.

Traditionally, nuclear structure problems are solved by following the time-dependent methods used in atomic structure with the Schrödinger approach. However, by going back to the more fundamental Dirac equation, many conceptual difficulties are found to be much easier to handle and better solutions are obtained, as in nuclear matter calculations or in intermediate-energy nucleon-nucleus scattering. Similar successes are found also in a variety of other problems in both nuclear structure and nuclear reaction. The calculations involved in solving four-component equations are more complicated than those in the two-component Schrodinger approach. In addition, we need more experience in handling certain aspects of the Dirac equation in a many-body setting. The results are, however, extremely encouraging and may lead to better understanding of some of the puzzles in nuclear physics. Another example is the use of sampling, or Monte Carlo, techniques in microscopic calculations. Hilbert space in a microscopic calculation can be extremely large and if a larger active space can be used, it can certainly reduce some of the uncertainties introduced by truncation and renormalization. Similar to many other types of problems, one can apply sampling techniques for certain investigations in large spaces and obtain meaningful results by carrying out only a small part of the actual work. The computer is well suited for doing this type of calculation, especially in view of the general trend toward parallel computing by making use of several central processing units at the same time. Furthermore, Monte Carlo techniques are used in a variety of other many-body problems, such as those in condensed matter physics, The advances made there can also be a great help in applying the method to nuclear physics problems. It is quite possible to start with a realistic nucleon-nucleon scattering potential and modify it so that it is appropriate for bound nucleons. The nuclear wave function obtained may be used to describe the nuclear state involved in an intermediate energy nucleon-nucleus scattering. Furthermore, the nucleon-nucleus interaction can also be derived from the same free nucleon-nucleon interaction potential used as the starting point for the nuclear wave function calculation. In each one of the steps, rigorous many-body problem techniques are available and can be applied. The calculation is a rather involved one. On the other hand, both the input nucleon-nucleon potential and the output nucleon-nucleus scattering can be checked directly with independent observations. The comparisons form the tests for many interesting questions, including many-body techniques and our understanding of how nucleons are modified inside the nuclear medium. Far more such large-scale calculations can be performed, especially in view of the higher precision experimental data that can he obtained these days. The use of numerical methods to solve research problems in astrophysics on a computer is a part of computational astrophysics. Numerical methods are used whenever the mathematical model describing an astrophysical system is too complex to solve analytically (with pencil and paper). So, application of numerical computational techniques for the solution research problems in nuclear astrophysics is inevitable. Let us have a short note of the essential numerical methods under the following subheadings:

1. **Computational versus Analytic Methods**

2. **Major Areas of Application of Computation in Nuclear Astrophysics**

3. **Numerical Methods**

4. **Stellar Structure Codes**

5. **Radiative Transfer Codes**

6. **N-body Codes**

7. **Codes for Astrophysical Fluid Dynamics**

8. **Relation to other fields.**

9. **Numerical Analysis**

10. **Computer Science**

11. **Relevant History**

10.2 Computational versus Analytic Methods

Solutions generated by numerical methods are generally only approximations to the exact solution of the underlying equations. However, much more complex systems of equations can be solved numerically than can be solved analytically. Thus, approximate solutions to the exact equations found by numerical methods often provide far more insight than exact solutions to approximate equations that can be solved analytically. For example, time-dependent numerical solutions of fluid flow in three dimensions can exhibit behavior which is not expected from one-dimensional analytic solutions to the steady-state (time independent) equations. The increase in computing power in the last few decades has meant that an increasingly larger share of problems in astrophysics can be solved on a desktop computer. However, the most computationally intensive problems in astrophysics are still limited by the memory and floating-point speed of the largest high-performance computer (HPC) systems available (e.g. computers on the top 500 list). The use of HPC is necessary to maximize the spatial, temporal, or frequency resolution of solutions, to include more physics, or when large parameter surveys are required to understand the statistics.

10.3 Major Areas of Application of CNA

Traditionally, the most important applications of Computation in Nuclear Astrophysics (CNA) have been found in the areas of:

10.3.1 Stellar structure and evolution

The internal structure and evolution of stars with different masses and chemical composition was largely mapped out in the 1960's using numerical methods to solve the equations of stellar structure. Today, the frontiers of research include calculating multidimensional stellar models of rapidly rotating stars, modeling the effects of hydro dynamical processes such as convection from first principles, and understanding how stars generate magnetic fields through dynamo processes.

10.3.2 Radiation transfer and stellar atmospheres

Without oversimplifying assumptions, computational methods are required to calculate the propagation of light through the outer layers of a star, including its interaction with matter through absorption, emission, and scattering of photons. The calculation of cross sections for the interaction of light with matter for astrophysically relevant ions is itself a challenging computational problem. The construction of such stellar atmosphere models share many challenges with radiation transfer problems in other systems such as planets, accretion disks, and the interstellar medium. Modern calculations improve the frequency resolution, include a better treatment of opacities, and can treat non-hydrostatic atmospheres.

10.3.3 Astrophysical fluid dynamics

The dynamics of most of the visible matter in the universe can be treated as a compressible fluid. Time-dependent and multidimensional solutions to the fluid equations, including the effects of gravitational, magnetic, and radiation fields, require numerical methods. A vast range of problems are addressed in this way, from convection and dynamo action in stellar and planetary interiors, to the formation of galaxies and the large scale structure of the universe.

10.3.4 Planetary, stellar and galactic-dynamics

It is well known that Newton's Laws of Motion for some number of particles N interacting through their mutual gravitational attraction do not in general have an analytic solution for N>2. Thus, to compute the orbits of planets in the solar system, or of stars in the Galaxy, numerical methods are required. The most challenging problems today include accurate integration of the orbits of the planets over the age of the solar system, studying the dynamics of globular clusters including the effect of stellar evolution and the formation of binaries, studying galaxy mergers and interaction, and computing structure formation in the universe through the gravitational clustering of collision less dark matter.

10.3.5 Numerical Methods:

The diverse set of mathematical models encountered in astrophysics means that a very wide range of numerical methods are necessary. These range from basic methods for linear algebra, nonlinear root finding, and ordinary differential equations (ODEs), to more complex methods for coupled partial differential equations (PDEs) in multi-dimensions, topics which span the entire content of reference books such as Numerical Recipes. However, there are several numerical methods used in astrophysics that deserve special mention, either because they have such wide use in astrophysics, or because astrophysicists have made significant contributions to their development. These methods are considered in the following subsections.

10.3.6 Stellar Structure Codes

The equations of stellar structure are a set of ODEs which define a classic two-point boundary value problem. Analytic solutions to an approximate system of equations exist in special cases (polytropes) but these are of limited application to real stars. Numerical solutions to the full system of equations were first computed using shooting methods, in which boundary conditions are guessed at the center and surface of the star, the equations are integrated outwards and inwards, and matching conditions are used at some interior point to select the full solution. Shooting methods are laborious and cumbersome; today modern stellar structure codes use relaxation schemes which find the solution to the finite-difference form of the stellar structure equations by finding the roots of coupled, non-linear equations at each mesh point. A good example of a public code that uses a relaxation scheme is the EZ-code, based on Eagleton's variable mesh method. The evolution of stars is then computed by computing stellar models at discrete time intervals, with the chemical composition of the star modified by nuclear reactions in the interior.

10.3.7 Radiative Transfer Codes

Calculating the emergent intensity from an astrophysical system requires solving a multidimensional integral-differential equation, along with level population equations to account for the interaction of the radiation with matter. In general the solution is a function of two angles, frequency, and time. Even in static, plane-parallel atmospheres, the problem is two-dimensional (one angle and frequency). However, the most challenging aspect of the problem is that scattering couples the solutions at different angles and frequencies. As in the stellar structure problem, relaxation schemes are used to solve the finite difference form of the transfer equations, although specialized iteration techniques are necessary to accelerate convergence. Monte, which adopt statistical techniques to approximate the solution by following the propagation of many photon packets, are becoming increasingly important. The problem of line-transfer in a moving atmosphere (stellar wind) is especially challenging, due to non-local coupling introduced by Doppler shifts in the spectrum.

10.3.8   N-body Codes

There are two tasks in an N-body code: integrating the equations of motion (pushing particles), and computing the gravitational acceleration of each particle. The former requires methods for integrating ODEs. Modern codes are based on a combination of high-order difference approximations (e.g. Hermite integrators), and symplectic methods (which have the important property of generating solutions that obey Liouville's Theorem, i.e. that preserve the volume of the solution in phase space). Symplectic methods are especially important for long term time integration, because they control the accumulation of truncation error in the solution. Calculation of the gravitational acceleration is challenging because the computational cost scales as N(N-1), where N is the number of particles. For small N, direct summation can be used. For moderate N (currently $N \sim 10^{5-6}$), special purpose hardware (e.g. GRAPE boards) can be used to accelerate the evaluation of $1/r^2$ necessary to compute the acceleration by direct summation. Finally, for large N (currently $N \geq 10^{9-10}$), three-methods are used to approximate the force from distant particles.

10.3.9   Codes for Astrophysical Fluid Dynamics

Solving the equations of compressible gas dynamics is a classic problem in numerical analysis which has application to many fields besides astrophysics. Thus, a large number of methods have been developed[1], with many impor-

---

[1]C. B. Laney (1998), Computational Gas dynamics, Cambridge University Press (ISBN: 0521570697)

tant contributions being made by astrophysicists. For solving the equations of compressible fluid dynamics, the most popular methods include finite-difference techniques (which require hyper-viscosity to smooth discontinuities), finite-volume methods (which often use a Riemann solver to compute upwind fluxes), operator split methods (which combine elements of both finite-differencing and finite-volume methods for different terms in the equations), central methods (which often use simple expressions for the fluxes, combined with high-order spatial interpolation), and particle methods such as smooth particle hydrodynamics (SPH), which integrates the motion of discrete particles to follow the flow[2]. A good technical review of many of these methods is given by Le Veque[3]. SPH is an example of a method developed largely to solve astrophysics problems, although many of the developments in other methods. The extension of finite-difference and finite-volume methods to magneto-hydrodynamics (MHD), to include the effects of magnetic fields on the dynamics of the fluid have also been motivated by astrophysics.

10.3.10   Relation to other fields:

Computational astrophysics is necessarily an inter-disciplinary subject, involving many aspects of astrophysics, numerical analysis and computer science. Few-body problems originating from the weak decays of hyper-nuclei (nuclei having one or more constituent nucleon(s) replaced by hyperons, e.g. Λ, Σ, Ω) in the stellar environment can also be included in computational astrophysics with brand name computational nuclear astrophysics.

10.3.11   Numerical Analysis

Numerical Analysis is a rigorous branch of mathematics concerned with the approximation of functions and integrals, and the approximation of solutions to algebraic, differential, and integral equations. It provides tools to analyze errors that arise from the approximations themselves (truncation error), and from the use of finite-precision arithmetic on a computer (round-off error). Convergence, consistency, and stability of numerical algorithms are all essential for their use in practical applications. Thus, the development of new numerical algorithms to solve problems in astrophysics is highly embedded in the tools of numerical analysis.

---

[2]J. J. Monaghan (1992), .Ann. Rev. Astron and Astrophysics, Vol. 30 (A93-25826 09-90), p. 543-574

[3]R. Le Veque (2002), Finite Volume Methods for Hyperbolic Problems, Cambridge University Press Olaf Sporns (2007) Complexity. Scholarpedia, 2(10):1623.

10.3.12 Computer Science

Let first note that computational science and computer science are not identical subject. The first one is the use of numerical methods to solve scientific problems while the latter is the study of computers and computation including various architectures of computers. Thus, computer science tends to be more orientated towards the theory of computers and computation, while scientific computation is concerned with the practical aspects of solving science problems. Still, there is a large degree of overlap between the fields, with many computer scientists engaged in developing tools for scientific computation, and many scientists engaged in software problems of interest to computer scientists. Examples of the overlap include the development of standards for parallel processing such as the Message Passing Interface (MPI) and Open MP, and development of parallel I/O file-systems such as Lustre.

10.3.13 Relevant History

Computational astrophysics has a long history, dating back to the numerical approximation of the motion of the moon and planets in order to compute ephemeris for tides and navigation. At first, the term computer referred to a person employed to calculate some quantity to some level of numerical accuracy. Electronic digital computers were introduced in World War II to break encryption codes and compute artillery range tables, however they were quickly adapted to other purposes after the war. In astrophysics, this included computing the first stellar structure models in the 1950s, and the development of numerical methods for fluid flow and shock-capturing at about the same time. Major advances in the decades that followed included the use of transistor instead of vacuum tubes, the development of compiled languages (Fortran in the 1950s, C in the 1970s), the adoption of the IEEE standards for the representation of floating point numbers and arithmetic (previously every manufacturer used their own pattern for bits) and the development of very large integrated circuits (VLSI), which allows the complexity of processor design that is enjoyed today. Modern trends in computer design are towards distributed memory parallel processors, with multiple cores capable of vector processing. At the same time as the development in hardware, progress in numerical analysis, computer science, and software engineering have resulted in standardized protocols for interprocessor communication across networks, object-orientated languages such as Fortran95 and C++, interpreted languages such as Java and Python, visualization tools such as OpenGL, and software engineering tools such as CVS and Subversion (SVN).

Exercise

1. What are the major areas of application of computation in nuclear astrophysics?

2. Compare the role of computational methods and analytical methods, in the study of nuclear astrophysics problems.

3. Explain how computation in nuclear astrophysics helps understandong the stellar structure and their evolution.

4. Which numerical codes are essential in the study of nuclear astrophysics? Write short notes on the N-body codes relevant to nuclear astrophysics.

5. Give a brief account of the major advancements in computational astrophysics.

References

Quoted in the text:

- C. B. Laney (1998), Computational Gas dynamics, Cambridge University Press (ISBN: 0521570697)

- J. J. Monaghan (1992), .Ann. Rev. Astron and Astrophysics, Vol. 30 (A93-25826 09-90), p. 543-574

- R. Le Veque (2002), Finite Volume Methods for Hyperbolic Problems, Cambridge University Press Olaf Sporns (2007) Complexity. Scholarpedia, 2(10):1623.

- Gregoire Nicolis and Catherine Rouvas-Nicolis (2007) Complex systems. Scholarpedia, 2(11):1473.

- James Meiss (2007) Dynamical systems. Scholarpedia, 2(2):1629.

- Mark Aronoff (2007) Language. Scholarpedia, 2(5):3175.

- Soren Bertil F. Dorch (2007) Magnetohydrodynamics. Scholarpedia, 2(4):2295.

- Kendall E. Atkinson (2007) Numerical analysis. Scholarpedia, 2(8):3163.

- Philip Holmes and Eric T. Shea-Brown (2006) Stability. Scholarpedia, 1(10):1838.

Chapter 11

Key challeges and future scopes

11.1   Introduction

Proper understanding of stellar burning processes possed many exciting challeges for physicsists across a wide range of nuclear physics topics, and the corresponding experimental programs are facing many technical drawbacks. For around forty years of research using beams of stable-isotopes, there are still reactions involved in quiescent burning whose cross sections are not known in the energy range of interest and, as a consequence, the astrophysical predictions are still have large degree of uncertainty for stellar systems that are, in principle, not very exotic, like the sun. Background sources in the laboratories are the major challenges for such measurements. Although underground facilities appears to offer a solution to overcome the experimental dificulties, they face other technical problems. The level of our knowledge of the properties of unstable nuclei and the reaction rates which are related to the explosive nucleosynthsis events and energy sources, reflect our level of understanding of the issues. The development of radioactive beams has made it possible to investigate some reactions involved in such explosive events. In traditional experimental approaches one uses stable-isotope beams meant for elastic and inelastic scattering, transfer reactions to obtain information on the properties of astrophysically important states of some radioactive nuclei. Indirect technique using, e.g., transfer reactions are particularly appealing for measurements with low-intensity radioactive beams. However, such approaches are model-dependent, and uncertainties related to the model will are reflected in their outcomes. In spite of the progress, many key reactions and key quantities, which are the most challenging ones, remain largely unknown. This is especially true for the r-process path. More intense and new radioactive beams and sophisticated detection systems are therefore the need of the time. On the other hand, nuclear theory has to be exploited to avoid misinterpretation of experimental data and a close collaboration with astronomical observations and astrophysical modeling is as essential as ever. In nuclear physics, after completing a nuclear structure calculation it is compared with experiment, and if necessary revisited taking the experimental results into consideration. While in nuclear astrophysics several additional steps are necessary. Since information on a broad range of nuclei is often required, development of an extrapolation or interpolation procedure may be necessary if the theoretical calculations are difficult or time consuming. The theory must then be validated by a comparison to an appropriate range of experimental data. If its predictions are not as accurate as is required, the theory must be "normalized" to experiment in an appropriate manner. Then, relevant cross sections or lifetimes must be calculated. This may involve straightforward computations of matrix elements or strengths. Or it may, for example, involve computations of inelastic neutrino scattering cross sections, or of radiative capture cross sections for isolated resonances, or the use of statistical approaches. Finally, these results must be expressed in terms of temperature dependent astrophysical rates. The complexity of the above task means that considerable thought must be given to decide when a new theory is sufficiently developed and sufficiently different from its predecessors to warrant revising the reaction rates. It also means that a coordinated effort must be made to address these problems since the theoretical strength at any single institution will be insufficient. Perhaps such a coordinated effort could help to ensure that the structure of the rate calculations and of the astrophysics programs that use them is such that improved theoretical and experimental information can be incorporated with relative ease. An important role of theory in nuclear astrophysics lies in providing guidance on the required measurements. Many astrophysical processes involve networks of reactions and decays; often it is far from clear whether a change in a particular nuclear property will have an impact in the astrophysical scenario. At other times, reactions occur under conditions of thermal equilibrium, and only the masses of nuclei and beta-decay strength are important. The importance of a given reaction may depend on the specific situation, for example on the mass of the star involved. Finally, it is clear that this process must be iterated, since changes in some rates may affect the importance of others. It will not be simple to provide guidance on the precision with which a given rate must be known, but it is essential to provide this guidance so that our experimental and theoretical efforts can proceed efficiently. A related issue is the need to ensure that the available data are quickly and conveniently accessible to those involved in using them directly for the generation of rates for astrophysical models, or as part of the evaluation process for theoretical calculations. Else much of the effort involved in making these difficult measurements will be washed out.

11.2 Key challenges

In spite of great advancement in nuclear astrophysics, there is still much more to be done to find the fruitful answers to many questions. Some of those includes:

- Our own planet earth and its rich biology depend on numbers of heavy elements synthesized during stellar evolution, violent explosive events like supernovae. What are the nuclear processes responsible for nucleosynthesis and when and where do they occur? What are the characteristics - temperature, density, and composition - of the nucleosynthesis sites?

- What is the explanation for the shortfall of neutrinos observed from our sun? Is the current discrepancy entirely the result of new physics beyond our standard theory of electroweak interactions, or does it represent, at least in part, some misunderstanding of the nuclear reactions that power our sun? What new technologies can nuclear physicists and others exploit to measure the entire spectrum of solar neutrinos?

- What drives the spectacular stellar explosions known as supernovae? What are the processes leading to the Type Ia supernovae used as standard candles in determining the acceleration of the universe? How do we correctly model core evolution in core-collapse supernovae? What determines whether a neutron star or black hole is left as a remnant? What is the site of the r-process? Do supernovae produce most of the short-lived gamma- emitting nucleus ^{26}Al that is found widely distributed in the galaxy? Detection of the various neutrino species emitted in such explosions may determine whether massive neutrinos play a central role in cosmology; what detectors might nuclear physicists construct for this purpose?

- What are the processes that led to the isotopic abundances observed in the presolar grains found in meteorites? Can these abundances serve to test models of s-process and r-process nucleosynthesis of the heavy elements in individual stars?

- What is the nature of the explosions that occur in accreting double-star systems? Can developments of present accretion models accurately describe novae, Type Ia supernovae, X-ray bursts, and X-ray pulsars? Do the proposed models for nucleosynthesis on the surface of accreting white dwarfs describe the abundance distributions in the ejecta of novae? How do the microscopic time scales of nuclear reactions affect the time scale of an X-ray burst?

- Does the big bang accurately describe the process that created the lightest elements? Can nuclear cross sections be measured with accuracy sufficient for big-bang calculations of the universal baryon density to provide a stringent cross check of the baryon density obtained from future measurements of the universal background radiation? Can results from the Relativistic Heavy Ion Collider (RHIC) on the nature of the quark/gluon to hadron phase transition shed light on the nature of in homogeneities in the universal density? Might there exist, in the nucleosynthesis expected from an inhomogeneous universe, evidence of early exotic states of high-temperature hadronic matter?

- Earth is bathed in a sea of cosmic radiation, much of it affected by nuclear processes occurring in our galaxy. How can further measurements of nuclear properties such as lifetimes, gamma-ray lines, and spallation cross sections help determine the origin of this radiation? How can we exploit unstable nuclei as cosmological clocks of past events in our galaxy? What is the origin of the highest energy cosmic rays?

- What exotic forms of nuclear matter exist at the extraordinary densities characteristic of neutron stars? What connections can be established between the observed properties of such stars - masses, radii, rotation rates, and electromagnetic emissions - and the behavior of nuclear matter under exotic conditions? How do the ashes of nuclear reactions involving proton-rich nuclei on a neutron-star surface affect its observable properties?

11.3   Demand for infrastructure expansion

An extensive program in nuclear astrophysics theory and experiment, will greatly advance our understanding of the cosmos. It will help strengthening observational and computational programs by providing the essential foundation necessary for the interpretation and simulation of new results. The infrastructure in nuclear astrophysics, especially in manpower and theoretical modeling, needs to be enhanced to meet these challenges. The field would benefit greatly from further co-ordination of its efforts and the increased educational opportunities for graduate students. Investment in specialized detection systems is also warranted. The many connections among astrophysical observables and astrophysical processes demand that nuclear astrophysicists take a broad view of their field. For example, evidence for neutrino oscillations follows from experiments with neutrino detectors for solar and atmospheric neutrinos. These oscillations may strongly affect element synthesis and the evolution of supernovae. Essential ingredients for the above facilities enhancement can be summarized as:

- A dynamic program of astrophysics studies at the new and upgraded radioactive ion beam facilities. Both fragmentation and ISOL(Isotope Separation On-Line) facilities are necessary to obtain the required information.

- Measurements of reaction cross sections with intense beams from low-energy stable-beam accelerators. Background reduction techniques are important. Measurements of reaction rates by indirect techniques and of spallation cross sections with higher energy stable-beam accelerators that provide energies from a few MeV to several GeV/nucleon.

- Operation and construction of solar neutrino detectors and the construction of advanced supernova neutrino detectors for studies of neutrino processes in supernovae.

- Utilization of neutron spallation sources for s- and r-process related experiments and possible measurements of neutrino-nucleus cross sections. Neutron beams from electron linacs and other sources, and gamma-ray beams from free-electron laser and synchrotron radiation sources are also important.

- Highly segmented high-efficiency gamma-ray detector arrays for the study of capture reactions and for the study of the structure of nuclei involved in the s-, r- and rp- processes.

- Large solid angle, good angular resolution, silicon detector arrays for the study of reactions with radioactive beams. These studies will provide masses, resonance energies, level parameters, spectroscopic strength, and decay widths for nuclei involved in astrophysical processes. These detectors require new developments in fast, high-density, low-noise electronics with a low cost per channel.

- High efficiency neutron detection for β-delayed neutron decay measurements. High efficiency and high angular resolution neutron detection for neutrons from charge exchange reactions to determine weak interaction rates in supernova processes and Coulomb-breakup experiments to determine neutron capture rates on radioactive nuclei.

- Magnetic spectrographs with large solid angle acceptances and mass separators with high beam rejection efficiency to provide the high selectivity and resolution required for many experiments. Advanced focal-plane detectors for high rate operation over a broad range of ion charge, velocity, and mass.

- Traps of ions and atoms for studies that require isolated atoms, including measurements of masses and studies of weak interaction properties.

- Dense targets (gas jet and gas cell targets, and liquid and solid targets at cryogenic temperatures), for measurements of small cross sections and for studies with radioactive beams. Development of targets of long-lived radioactive nuclei, especially for studies of s- and p-process nuclei. The expense and availability of rare and separated stable isotopes needs to be addressed.

- The infrastructure in nuclear astrophysics, especially in manpower and theoretical modeling, should be strengthened. Additional support for graduate and postdoctoral education is needed. Enhanced and structured collaborations among institutions with different capabilities would contribute greatly to the advance of the field.

- Nuclear structure and reaction theory is an essential ingredient of nuclear astrophysics. A major effort is required to connect theoretical calculations to experimental data, to determine astrophysical reaction rates, and to summarize this knowledge in a form useful for astrophysical calculations. A related effort must ensure that the extant experimental data are available to nuclear theorists and astrophysicists for calculation of reaction rates or validation of theoretical models.

- Multidimensional simulations will be an important bridge between nuclear data and astrophysical observations. Group efforts involving both nuclear physicists and astrophysicists are probably the best approach to the simulation of important and complex phenomena such as supernovae. Major computing resources (tera- and ultimately peta-flop scale) will be required.

11.4 Progress Scenario

While strong experimental radioactive beam programs in nuclear astrophysics are presently being pursued at all major research facilities, new initiatives to improve beam intensities for already available radioactive beams and to provide opportunities for beam developments closer or even at the drip lines dominate the discussions worldwide. ISAC at TRIUMF (Canada), REX ISOLDE at CERN and SPIRAL at GANIL (France) are starting to operate as the most recently completed radioactive beam facilities, while new developments on a larger scale are planned for the future. The upgrade of the RIKEN accelerator facility (JAPAN) to a radioactive beam factory RIBF (radioactive ion beam factory) is projected for completion and operation in 2007. The project features a superconducting ring cyclotron and a new projectile fragment separator system to provide high energy radioactive beams. In addition in-flight fission of uranium will be used to produce very neutron rich medium mass nuclei whose study will be important for the interpretation of the r-process. Activities in Europe presently focus on the proposed implementation of new technological developments at GSI Helmholtz Centre for Heavy Ion Research in Germany which are centered around the construction of a fast cycling superconducting double-ring synchrotron with a system of storage rings for beam collection and beam cooling. Completion of this project will lead to an unprecedented range of experimental opportunities with fast radioactive beams originating from fragment separation. Nuclear Astrophysics far off the limit of stability is being envisioned as one of the prime research area pursued at the new facility. The projected experiments will most likely focus on nuclear structure measurements - half-lives, masses, decay properties - along the rp-process and the r-process path and may extend to the measurement of p-process reactions using the Coulomb dissociation method. In addition to pure decay or ground state properties on the one hand and reaction cross section measurements on the other hand, there is the chance to investigate also other features like the density of excited states, giant (E1,M1) resonance properties as a function of distance from stability or the Gamow-Teller strength distribution for unstable nuclei. Such measurements can help to constrain the individual ingredients

of theoretical predictions for strong, electromagnetic and weak interaction
cross sections. Yet, in addition to reaction properties which were at the
center of this discussion relevant to astrophysical issues of radioactive ion
beams, the GSI facility will also provide unique opportunities for probing
experimentally quark gluon plasma matter close to the conditions of neutron
star matter. This aspect is also of significant interest for the nuclear astrophysics community since it may provide novel experimental opportunities
to test not only quark-gluon plasma matter at conditions similar to those
in the center of a neutron star but also the transition to normal nuclear
matter as anticipated for the lower layers of the neutron star crust. This
has an impact for a number of environments discussed in this description
like type II supernovae, neutron star mergers as well as X-ray bursts. The
most ambitious future project yet is RIA, the Rare Isotope Accelerator, a
US project which is presently awaiting approval by the US Department of
Energy. RIA is centered around a 1.4 GeV multi-beam LINAC driver accelerator capable of providing beams from protons to uranium at energies of at
least 400 MeV per nucleon, with beam power in excess of 100 kW. With this
flexibility, the production reaction can be chosen to optimize the yield of a
desired isotope. In comparison to the two main competing in-flight facilities,
the Radioactive Ion Beam Factory at RIKEN and the next-generation GSI
facility, RIA has two advantages. RIA seeks to combine both technological
methods of producing radioactive beams by providing low energy ISOL separated beams and fast energy fragment separated radioactive beams. RIA's
capability for post acceleration of ISOL based radioactive particles is not
included in either of the other two projects. This dual-concept approach
allows to develop maximum intensity beams over the entire nuclear chart
using the most advantageous production mode, will allow a wider range of
experimental studies. These will expand to the direct measurement of nuclear reactions at astrophysical energies beyond the measurements of nuclear
structure characteristics relevance for astrophysics which is the main focus
for experiments with fragment separated high energy beams. Progress in
nuclear astrophysics is truly multi-disciplinary and multi-national: advances
in astronomical instrumentation are generating ever-improving imaging and
spectroscopy over both wider and deeper fields of view; high-precision analytical equipment increasingly allows us to measure minuscule amounts of
rare isotopes in terrestrial rocks and meteorites; and sophisticated accelerator systems and detectors potentially provide the tools to probe unusual,
transient nuclei and reactions. All this research is brought together and
complemented by theoretical work, often requiring high-performance computing. Further progress can be achieved through a greater synthesis of

research efforts across the disciplines, based on focused collaboration between different expert groups. The education of young scientists needs to be enriched with specialized interdisciplinary courses and workshops. Along this line, some initiatives have been set up by the European Science Foundation, such as Euro Genesis (http://www.esf.org), which addresses the origin of the elements and the nuclear history of the Universe; and Comp Star (http://compstar-esf.org), focusing on compact stars and supernovae.

11.4.1 Available Data Resources

Since the last few-decades there have been major advances in evaluating and making publicly available astrophysical and nuclear data for nuclear astrophysics. Major efforts that make results publicly available and address specific needs in nuclear astrophysics are summarized below:

- Big Bang Online: http://bigbangonline.org is a Cloud computing system that provides codes and data related to Big Bang nucleosynthesis.

- Cococubed: http://cococubed.asu.edu/code_pages/codes.shtml provides a set of useful Fortran codes for nuclear astrophysics.

- JINA: http://www.jinaweb.org The Joint Institute for Nuclear Astrophysics (JINA) provides a frequently updated database of currently recommended stellar reaction rates (JINA reaclib), a public R-matrix code (AZURE), and a virtual journal that identifies literature with new data for nuclear astrophysics.

- KADoNiS http://www.kadonis.org provides an occasionally updated and well documented data base of evaluated s- and p-process stellar reaction rates.

- Livermore: http://adg.llnl.gov/Research/RRSN/ provides a website with links to a broad range of nuclear and astrophysical data for use in astrophysical models.

- MINBAR: https://burst.sci.monash.edu/wiki /index.php? n = MINBAR.Home The Multi Instrument Burst Archive will provide data for more than 6000 X-ray bursts that are consistently analyzed.

- MESA: http://mesa.sourceforge.net is a modern 1D stellar evolution code that is open source, modular, takes advantage of modern computational techniques, and is well supported. It includes nuclear reaction networks.

- NACRE: http://pntpm.ulb.ac.be/Nacre/nacre_d.htm provides a set of evaluated rates for reactions with stable nuclei from 1999.

- NETGEN and BRUSLIB: http://www.astro.ulb.ac.be/pmwiki /IAA /Databases provides databases for reaction rates and tools to create tailored reaction networks.

- NNDC http://www.nndc.bnl.gov/astro/ The National Nuclear Data Center (NNDC) provides stellar neutron capture rates evaluated based on the NNDC evaluated nuclear data.

- NuGrid http://www.nugridstars.org makes available a number of tools and data products for nucleosynthesis calculations, including a virtual box based nova model.

- nucastro.org http://nucastro.org provides various nuclear astrophysics data sets based on theoretical calculations with the Hauser-Feshbach approach.

- nucastrodata.org http://nucastrodata.org provides a computational infrastructure for nuclear astrophysics, including tools to evaluate and calculate reaction rates and to carry out reaction network calculations.

- SAGA: http://saga.sci.hokudai.ac.jp/wiki/doku.php provides a database for stellar abundances.

- STARLIB: http://starlib.physics.unc.edu provides stellar reaction rates derived with a novel Monte Carlo method for estimating experimentally-based reaction rates, and associated uncertainties, in a statistically meaningful manner.

- Webnucleo: http://nucleo.ces.clemson.edu provides a variety of public codes for nuclear astrophysics, including the reaction network code libnucnet. XML is used as data format, and tools are available for converting and manipulating data.

11.4.2   Future Data Developments

Continuity

Continuous support for existing database efforts in nuclear astrophysics is critical. Without long term continuity databases become quickly outdated, and new data that are often obtained using significant resources cannot be used in astrophysical calculations. Continuous support is therefore essential for rapid progress in the field, for taking advantage of nuclear and observational data, and to ensure researchers are not reaching the wrong conclusions because of the use of outdated nuclear data.

Evaluation

A community-wide effort is needed to identify and evaluate important stellar reactions. Researchers should reach out to the astrophysics modeling community to provide them with needed input as well and to the nuclear data community for their expertise in evaluations. The community must develop a set of best practices for rate evaluations, and communicate these widely together with the necessary data, tools, and codes. This must include the evaluation and proper determination of uncertainties. A series of workshops may be a good approach to achieve this goal.

Transparency

While already a number of open source nuclear astrophysics codes exist, it will be important in the future to expand the number of open source codes. Open source has many advantages, including broader use (advancing the science more rapidly) and community input and contributions on code improvements. On the other hand, concerns about return of investment for funds and efforts by individual institutions and researchers, and concerns about ongoing code support have to be addressed.

Ease of Use

Ease of use could be improved for many public nuclear astrophysics codes. Possibilities include cloud computing or virtual boxes to avoid compatibility, update, version, backup, or cyber security issues. GUIs could be customized for different users. No single data format will work for all the diverse phenomena in nuclear astrophysics but robust database storage with custom graphical user interfaces are an excellent solution for many cases. The use of XML as a standardized but flexible format is being explored. By choosing a standard format like XML, it would be possible to take advantage of freely available tools (such as XSLT in the case of XML) for converting the data to other formats.

Distribution and Hubs

Multiple distribution sites are currently quite effective in satisfying diverse user needs. However, there are also drawbacks from such an approach. It would therefore be interesting to explore a unifying HUB Zero-based approach for nuclear astrophysics. Such an approach has been successful for other communities, for example nano- technology (nanoHub.org). A hub could be created together with the broader nuclear physics community. HUB Zero is an open source system that enables a wide variety of content (lectures, animations, codes, databases, tutorials, and so forth) to be put online. A set of well-developed and easy-to-use protocols allows users to install executable versions of their own research codes (in Fortran, C, C++, etc.) on the HUB without having to write their own web software. HUB Zero middleware then allows these codes to be executed on the full complement of computer resources available on the Open Science Grid, and standardized graphical interfaces allow the results to be visualized with a variety of plotting routines. By streamlining the installation of codes on the web, a HUB allows a community to leverage the efforts of a much larger fraction of its membership to develop shared resources than if those members developed and maintained their own separate web sites. Furthermore, by adopting a HUB Zero-based approach, the nuclear astrophysics community could also benefit from the considerable experience of other communities that have their own HUBs.

Cloud Computing Opportunities

"Cloud computing" services may also be a transformative vision for the future of the field. This opens up many possibilities including: having a digital assistant who automatically collects relevant masses, level schemes, references; a way for experts to easily upload supplemental information for your evaluations; having all major databases just one mouse click away; having an evaluation template automatically filled out for you; running analysis and application codes without compatibility, updates, backups, or cyber security issues; designing custom views of datasets from a variety of visualization tools; having a "virtual expert" online 24/7 to consult with questions; sharing your large data sets easily with colleagues; easily uploading your evaluation and visually tracking its progress for reviews, revisions, and acceptance; using a pipeline to process your evaluated data for use in simulations codes; running and visualizing these simulations, then sharing the results with colleagues

11.5 Present and Future Projects & their Benefits

This section introduces some international projects that will benefit nuclear astrophysics development as well as to the society.

11.5.1 Astronomical observatories

Improved (multi-object) spectrometers at all wavelengths will provide extensive information on the origin of the elements:

- Gaia - the ESA Gaia spacecraft, launched in December 2013 is surveying a billion stars in our Galaxy, and will provide spectroscopic information on stellar structure and evolution.

- LAMOST -the recently commissioned Large Sky Area Multi-Object Fibre Spectroscopic Telescope in China is carrying out a five-year spectroscopic survey of 10 million stars in the Milky Way.

- ALMA - the international Atacama Large Millimetre Array, under construction in Chile, will analyze the composition of gas clouds where stars are born.

- e-ELT - the 39-metre European Extremely Large Telescope, also to be built in Chile, will be ready in the early 2020s, will study in spectroscopic detail the birth and evolution of stars and planets.

- INTEGRAL carries the only current telescope which can measure the variety of gamma-rays from cosmic radioactive nuclei; this ESA mission was launched in 2002 and may continue into the next decade.

- JWST - the 6.5-metre James Webb Space Telescope, to be launched in 2018, will complement terrestrial telescopes from space, to reach the earliest stellar generations.

- NuSTAR - the NASA Nuclear Spectroscopic Telescope Array 2012 mission maps in X-rays newly synthesized titanium-44 in the debris of nearby young supernovae.

- ASTRO-H and ATHENA - are Japanese and European X-ray spectroscopy projects respectively, planned for launch between 2015 and 2028, which will measure the hot interstellar gas related to regions where nucleosynthesis happens.

- Laboratory analysis New generations of instruments capable of probing material with nanometer-resolution are being developed to measure pre-solar grains more efficiently. Lasers are used to measure the precise composition of selected isotopes in samples.

- Accelerator based research A new generation of facilities will enable experiments probing nucleosynthesis, and the extreme conditions in which they occur, to be carried out:

- Radioactive ion beams – several European nuclear research laboratories, including GSI (FAIR) in Germany, CERN (HIE-ISOLDE) in Switzerland, EURISOL (several European sites hope to host this next-generation experiment), and Legnaro (SPES) in Italy, open a new era with next-generation facilities: These create nuclei never made before, and allow their properties to be determined - essential for understanding nuclear reactions in cosmic explosions.

- FAIR - the Facility for Antiproton and Ion Research will host a range of projects enabling new research on the properties of dense nuclear matter and atomic structure in extreme electromagnetic fields, advancing our understanding of neutron stars and of plasma conditions inside stars and planets.

- LUNA-MV - the LUNA Collaboration is developing a more powerful accelerator in the Gran Sasso underground laboratory to measure accurately the very slow reactions associated with hydrogen burning.

- Neutron sources - more intense neutron beams are being developed, such as the European Spallation Source and FRANZ, which will be capable of measuring neutron-capture reactions.

- ELI -one component of the planned European laser facility, the Extreme Light Infrastructure, will be built in Romania and dedicated to nuclear physics experiments.

11.5.2   Benefit to theory and computation

Advances in computers are the key to developments of the complex theories of nuclear structure, supernova explosions, and cosmic evolution of the abundance of the elements: superfast calculations enable theories and their parameters to be widely explored, and to guide experiments and compare their results with theoretical predictions. Also support for databases of thousands of nuclei and their reactions, and advanced graphics-processing to visualize complex data or theories, are key drivers of the field.

11.5.3 Benefit to the society

The technology and skills developed for nuclear astrophysics research are helping to find solutions to many of the challenges our society faces today. Detectors developed for nuclear physics have been adapted for medical imaging and diagnosis:

- MRI: (magnetic resonance imaging), PET (positron emission tomography) and CT (computer-aided tomography). Beams of atomic nuclei are also used to destroy cancerous tumors, and radiotherapy based on injections of radioactive substances is a part of cancer treatment. Ionizing and nuclear radiation is also used to sterilize medical equipment, household items and food, by utilizing their lethal effects on microbes.

- Computing and information processing: The computational tools developed by researchers modeling stellar interiors and supernovae explosions have inspired a variety of computing methods used in, for example, medical imaging and engineering, or simply to provide faster internet support and faster processors.

- Energy: Energy supply is one of the biggest challenges for the next decades, and many countries will continue to rely on nuclear power. The techniques of nuclear astrophysics are used to improve the efficiency and safety of nuclear reactors, and technology is now available to 'clean' the radioactive waste so far generated. In the future, reactors employing fusion reactions like those in the Sun will generate safe nuclear energy.

- Environment and analysis: The evolving global climate and our effect on it is of great importance. Technologies from astrophysics and nuclear physics have improved tools for remote-sensing. Radioactivity from both natural events, such as volcanoes, and manmade sources are monitored. Measurements of trace isotopes can track subsurface water flows. Detecting minute amounts of characteristic isotopes was pioneered in nuclear astrophysics laboratories, and now allows us to identify art forgeries, determine the age of artifacts and materials, analyze geological samples, and monitor environmental pollution. Security has been improved by detectors and analytical techniques from nuclear astrophysics that can track sensitive materials at national borders, and scan cargo and baggage for explosives, radioactive or fissile materials. Short-lived radioisotopes can probe manufacturing processes and analyze product performance, for example, wear and tear in engine components..

- A skilled workspace: Understanding the puzzles of the Universe attracts talented people to study physics, and other numerate disciplines. This training produces highly-skilled individuals with the analytical and technical abilities needed by industry and government to solve the problems faced by society in the present days.

Exercise

1. Briefly address the major challenges in nuclear astrophysics.

2. Discuss the direct and indirect benefit of the technological developments for nuclear astrophysics, to the society.

3. Give a brief account of the present and future projects in nuclear astrophysics.

Index

abundance, 61
accelerator, 13
accretion, 122

beta-decay, 25, 26,
    89, 121
binary star, 135,
    231, 232
binding energy, 18
black hole, 12
bremsstrahlung, 133
brown dwarf, 96

Chandrasekhar
    mass, 50
CNO cycle, 99, 101
collisions, 40, 49
Compound nucleus,
    21
COMPTEL, 235
core collapse, 2,
    167, 194
cosmology, 12
cross section, 22,
    32, 33

dark matter, 14
deformation, 57
degeneracy, 194,
    195, 197,
    233
drip-line, 221
dwarf, 14

endothermic, 22
entrance channel,
    21
entropy, 202
evolution, 2, 63, 65,
    67
excitation, 106, 140
exothermic, 22

FRDM, 220

Gamma-decay, 27

half-life, 14, 17

ignition, 115, 180,
    183, 231
ionization, 33
ISAC, 137
ISOL, 136, 137, 271

LNGS, 33

mean-life, 40, 84

neutrino capture,
    114, 117,
    223

neutron capture, 13,
    77, 81, 84
NSE, 205
nuclar fusion, 16, 18
nucleosynthesis, 11,
    13

primordial, 70

Q-value, 21
quiescent, 123

r-process, 2, 64, 66
radioactivity, 13, 14
reaction network,
    114, 118,
    200
reaction rate, 11,
    36, 74, 76,
    77
reactor, 13
resolution, 125
resonances, 37
Roche Lobe, 232
RPA, 209

S-factor, 34, 134,
    152
s-process, 65, 84,
    86, 87, 108

scattering, 21, 22, 32

simulation, 135, 139, 169, 197

spallation, 133, 136, 223

stability, 23

star merger, 2, 113, 226, 227

stardust, 54

stoichiometry, 122

supernova, 14, 63

transition rate, 32

tunneling, 29

universe, 15

weak-interaction, 32

white dwarf, 53

X-ray bursts, 2, 122

Printed and bound by CPI Group (UK) Ltd, Croydon, CR0 4YY

17/10/2024

01775694-0011